Guidelines for
Baseline

Ecological

Assessment

Guidelines for
Baseline

Ecological

Assessment

INSTITUTE OF
ENVIRONMENTAL
ASSESSMENT

E & FN SPON

An Imprint of Chapman & Hall

London • Glasgow • Weinheim • New York • Tokyo • Melbourne • Madras

Published by E & FN Spon, an imprint of Chapman & Hall,
2-6 Boundary Row, London SE1 8HN, UK

Chapman & Hall, 2-6 Boundary Row, London SE1 8HN, UK

Blackie Academic & Professional, Wester Cleddens Road, Bishopbriggs, Glasgow G64 2NZ, UK

Chapman & Hall GmbH, Pappelallee 3, 69469 Weinheim, Germany

Chapman & Hall USA., One Penn Plaza, 41st Floor, New York, NY10119, USA

Chapman & Hall Japan, ITP-Japan, Kyowa Building, 3F, 2-2-1 Hirakawacho, Chiyoda-ku, Tokyo 102, Japan

Chapman & Hall Australia, Thomas Nelson Australia, 102 Dodds Street, South Melbourne, Victoria 3205, Australia

Chapman & Hall India, R. Seshadri, 32 Second Main Road, CIT East, Madras 600 035, India

First edition 1995

© 1995 Institute of Environmental Assessment

Printed and bound in Hong Kong

ISBN 0 419 20510 1

A Catalogue record for this book is available from the British Library

Contents

Foreword

Environmental assessment (EA) is now a well established process in the United Kingdom, but there is a growing need for specific and detailed guidelines into the different aspects of EA. The Institute of Environmental Assessment has been instrumental in the development of the EA process in the UK and has set itself the target of providing a range of guidelines on reasonable best practice in this field. These guidelines, concerning ecological aspects of environmental assessment, form part of a growing series being produced by the Institute.

Ecological issues are often poorly attempted in EA and the guidelines will be an important step to improving the quality of environmental assessment in the UK. The publication also provides a unique source of information on survey methods, for a wide range of biological groups, that can be used in ecological assessment in the UK.

This document is the culmination of a period of dedicated work by a team of professional ecologists and environmental scientists from across industry, consultancies, statutory organisations, local authorities, non-governmental organisations and universities. The Institute and the working group chair would like to thank them, and many others who provided specialist advice, for their invaluable contributions.

The responses to drafts of the guidelines have been very positive and I believe the guidelines will be a valuable reference for industry, local planners and consultants alike.

National Power is very pleased to have been involved in sponsoring this important and timely document.

John Baker
Chairman
National Power PLC

National Power

Preface

These guidelines were commissioned by the Institute of Environmental Assessment and undertaken by a Working Party comprising representatives from local authorities, non-governmental organisations, universities, consultants, statutory agencies and developers.

A Working Party chaired by Dr Martin Marais, National Power Plc and previously by Dr Brian O'Connor, Joint Nature Conservation Committee met on eight occasions, between 1991 and 1994, to consider the various drafts and to input ideas. It is intended that this document will be periodically reviewed and updated in the light of evolving practice and legislation.

In alphabetical order the Working Party members were:

Dr Christopher Betts	Christopher Betts Consultancy
Stephen Brooks	Natural History Museum
Dr Peter Bulleid	Barton Willmore Partnership Ltd
Dr Tim Coles	Rapporteur for IEA
Jaqueline Fisher	Jaqueline Fisher Associates
Dr Margaret Hill	Environmental Advisory Unit Ltd
Richard Knightbridge	Land Use Consultants
Tom Langton	Herptofauna Consultants International
Dr Paul Logan	National Rivers Authority
Dr Martin Marais	National Power Plc
Mike Oxford	Avon County Council
Margaret Palmer	Joint Nature Conservation Committee
Dave Pritchard	Royal Society for the Protection of Birds
Dr Tim Reed	Joint Nature Conservation Committee
Dr Geoff Ricks	RPS Cairns
Dr Richard Snowdon	Sir William Halcrow & Partners
Amanda Smith/Anne Goodall	Ecosurveys Ltd (Editors)
Brian Smith	English Nature
David Stubbs	Institute of Ecology and Environmental Management
Dr Jo Treweek	English Nature
Dr Peter Wathern	University of Wales, Aberystwyth
Dr Colin Welch	Institute of Terrestrial Ecology

The production of the guidelines was funded by National Power Plc. The Working Party would like to thank the following, who have provided valuable input on previous drafts: Dr John Box (English Nature); Dr Steven Crute (Private Consultant); David John (Natural History Museum); Deborah Nissenbaum (Environmental Advisory Unit Ltd); Dr Nick Davidson (Joint Nature Conservation Committee); Dr Max Hooper (Institute of Terrestrial Ecology); Dr Mary Mitchell (Independent Advisor); Steve Muddiman (Environmental Advisory Unit Ltd); Peter Nelson (Land Use Consultants);and Jenny North (University of Wales, Aberystwyth).

Any views expressed in the guidelines are the views of those who have contributed to the guidelines and do not necessarily reflect the views of the organisations which they represent.

Acknowledgements

The Institute of Environment Assessment gratefully acknowledges the illustrations contributed by the following organisations: Anne Goodall (cover, p4, p40); Biological Records Centre, ITE Monks Wood (p22); Ecosurveys Ltd (p14, p52, p58); English Nature (p25, p32); John Feltwell/Wildlife Matters (p11, p43, p48, p49); Lincolnshire Trust for Nature Conservation and National Trust (p22); Tim Smith (p64).

Summary

These guidelines have been compiled by an Institute Working Party, comprising representatives from local authorities, universities, non-governmental organisations, consultants, statutory agencies and developers. They offer general guidance on the extent to which baseline ecological information should be presented in various situations. The guidelines also provide a unique source of information on survey methods, for a wide range of biological groups that can be used in ecological assessment throughout the UK. This document offers guidance as to current best practice and will be valuable to project managers, competent authorities and professional ecologists alike.

It is of paramount importance that individuals only undertake survey work that is within their area of competence. The professional competence of ecologists to be engaged on a project may be gauged by their attainment of a combination of credentials including academic qualifications; proven track record; membership of an appropriate professional body; fulfilment of continuing professional development requirements; internal training programmes; and identification qualifications.

The guidelines represent a consensus on the level of baseline data required to assess adequately the ecological impacts of a proposed development. Prior to starting the assessment process it is essential that the scope of the Environmental Assessment is correctly defined. Scoping the ecological elements of these studies is, however, an iterative process and involves some initial work in order to determine whether the development will significantly impact any important ecological features. It is also important to collect any existing ecological data on the area to be affected and consult with various statutory and voluntary conservation agencies. Recognised sites of nature conservation interest (statutory and non-statutory) which are within or near areas likely to be directly or indirectly impacted should be identified within an environmental statement and the main reasons for their designation outlined. A minimum 2km radius around the site should be considered although in some cases impacts can occur further afield.

Several baseline survey techniques are described for most habitat types. The use of the Phase 1 survey methodology is recommended. The maps should be accompanied by extended target notes, which identify and provide further information on habitat features of particular value to different ecological groups such as plants, fungi, lichens, mosses and birds. Habitats along a river can be described using the NRA river corridor survey method and for subtidal habitats on hard coasts the SEASEARCH method would be appropriate.

The guidelines describe the criteria for triggering more detailed Phase 2 studies for the following major groups of living organisms: vegetation; birds; mammals; amphibians & reptiles; fish; terrestrial & aquatic invertebrates. The final section is devoted to marine and estuarine habitats and organisms. Areas of impact in these habitats are potentially large and repeated sampling is often required because of the variation between seasons and tidal cycles. The appendices provide useful reference information on statutory consultees, UK protected species, licensing procedures, Annex 1 habitats (EC Habitats Directive) found in the UK as well as a list of professional institutes relevant to ecologists.

In view of the broad range of specialisms inherent in ecology these guidelines have been produced after an extensive consultation period. The guidance offered as to current best practice will make it valuable to everyone who is involved in ecological surveying at every level.

PART ONE

Introduction

Aims of the guidelines

1.1   This guidance note has been prepared as one of a series of documents published by the Institute of Environmental Assessment (IEA) to establish good practice for carrying out various aspects of Environmental Assessment (EA) in the UK.   Environmental Assessment is an environmental management tool which has been in use on an international basis since 1970. It is a process by which the identification, prediction and evaluation of the key impacts of a development are undertaken and the information gathered is used to improve the design of the project and to inform the decision making process. The process is illustrated in Appendix 1. EA was formally introduced in the UK through the EC Directive of 1985 (CEC, 1985) and was implemented in 1988 through a series of statutory instruments known as the EA regulations.

1.2   These guidelines concentrate on outlining best practice for describing and evaluating the **ecological baseline** of an EA. The guidelines should not be regarded as exhaustive or as providing definitive advice but merely as general guidance on the extent to which ecological information should be presented in various situations.

1.3   The guidelines are needed:

* to guide ecologists in their work and to promote best practice.
* to inform project managers and clients about the role of ecology in EA and help them to set appropriate briefs and budgets.
* to guide the determining authorities in assessing ecological statements accompanying EAs.

Scope of the guidelines

1.4   These guidelines, have been prepared by an Institute of Environmental Assessment Working Group (see Preface) and have been subject to an extensive public consultation phase. They represent a consensus on the level of baseline data required to assess adequately the ecological impacts of proposed developments. After the **Introduction, Part Two** explains the importance of consultation and scoping, and outlines some of the general issues that should be

4

considered when planning and undertaking more detailed ecological studies. **Part Three** presents the specific criteria which would indicate the need to present more detailed data on various ecological groups, together with recommended survey methods and guidance on evaluating the collated baseline data.

1.5   Unfortunately, due to constraints on time and information, it has not been possible to include specific criteria and survey methods for fungi. However, fungi play an important role in the ecosystem and it may need to be considered in an ecological assessment. Therefore it is intended to include a section on fungi in a future edition of the guidelines. Finally, **Part Four** covers the requirements for more detailed surveys in marine and estuarine systems.

1.6   The ecological impacts of proposed developments may involve habitat loss or modification and a potentially wide range of other impacts which will vary from project to project and according to location. Due to the wide variation of both habitats and of the proposed developments which may affect them, these guidelines are designed to be flexible. Much effort has gone into ensuring that the methods outlined are the best available for studying particular species and groups. However, each assessment will be unique and the method chosen to characterise baseline ecological conditions is likely to be tailored to the site and the impacts expected. Therefore, it is for the ecologist(s) conducting baseline surveys to make their own professional assessment in the choice of methods to be used. These guidelines should be seen as an aid to practitioners in deciding the level of detail required to characterise baseline conditions adequately.

False colour photography can be used to identify stressed vegetation. The dark band at the top is a well watered golf course, the pale area in the middle is a heathland nature reserve and the line of white oaks towards the bottom of the picture are under water stress. The recent conifer planting has greatly affected the water regime in the vicinity

1.7   The first steps in an ecological assessment are to visit the proposed development site and its surrounds, collect any existing ecological data on the area to be affected and consult with various statutory and voluntary conservation agencies.

1.8   However, it is normally the case that insufficient data exist upon which a judgement about ecological impact can be based. Therefore, the main problem facing the ecologist is how much data will need to be gathered to supplement pre-existing information and to assess whether a significant impact on a particular taxonomic group or ecosystem will occur.

1.9   The issues covered by these guidelines are, of course, only part of the Environmental Assessment process. The guidelines do not address how to assess and evaluate the ecological impacts of alternative proposals, how to assess changes in the baseline conditions that would occur in the absence of the project proceeding, how to predict and quantify ecological impacts, how to mitigate these impacts, how to assess the significance of residual (after mitigation) changes brought about by the development or to suggest ecological monitoring requirements. Guidance on some of these issues has been or is about to be published (Forbes and Heath, 1990; Walsh *et al.* 1991; Box and Forbes, 1992; Department of Trade, 1993; English Nature, in prep. Department of the Environment, in prep.) and will also be addressed by future IEA working groups.

1.10   Whilst these guidelines have been prepared as advice on reasonable best practice for part of the Environmental Assessment process, the same principles may be used for determining the level of information required for other ecological reports (eg environmental appraisal reports accompanying planning applications, ecological elements of CIMAH assessments, IPC applications, BS7750 and Eco-Management and Audit Regulation Environmental Effects Registers, and licences and consents under the Water Resources Act 1991).

Summary

- The guidelines have been prepared by an IEA working group and published after extensive consultation with the various interested parties in the Environmental Assessment process.

- The guidelines represent a consensus viewpoint as to the level of baseline data that is required to adequately assess the ecological impacts of proposed developments. Different aspects of the Environmental Assessment process are covered by other publications, and will also be addressed by future IEA guidelines on these aspects.

- The ecological criteria set out in the guidelines could also be used to help in defining the level of information sufficient to characterise ecological conditions for other types of ecological appraisals.

PART TWO

Principles and good practice of
ecological assessment

Defining the Scope of an Ecological Study

Introduction

2.1   Correctly defining the scope of an EA prior to starting the assessment process is essential to the production of a good quality Environmental Statement. Scoping is responsible for focusing on a subset of key issues and potential impacts, some of which will be ecological impacts.

2.2   For many types of impact (eg noise and visual impacts) an initial visit to the site, consultation with interested parties and a full description of the proposed development enables the scope of the study for these issues to be determined. With ecological impacts, however, some initial work (eg data collection, habitat assessment) in addition to the above stages often needs to be done in order to identify the important issues on which to focus the study. Determining the scope of the ecological aspects of the EA is thus more of an iterative process - a certain level of work needs to be done in order to determine:

a)   whether or not there are issues of ecological impor-
                tance for the site.

b)   where significant impacts are predicted, whether
                there are suffcient data available on the various
                ecological communities which may be affected to
                assess the magnitude and significance of those
                impacts or whether as part of the assessment process
                additional survey information will be required.

2.3   This section defines the work necessary to determine the above points and thereby scope the ecological elements of the EA. This process is illustrated in Figure 1 (pg.14). It therefore represents the level of ecological study which should be undertaken for all EAs and comprises the following elements:

•   consultation and the gathering of relevant existing ecological
        data for the affected site and its surrounds.

•   a site visit and preparation of a relevant type of Phase 1 habitat map identifying any areas of importance for floral or faunal communities.

2.4   All ecological field surveys must be carried out by appropriately qualified ecologists, with relevant field experience of the survey methods being used and of the species or habitats under study. In view of the broad range of specialisms inherent in ecology, it is of paramount importance that individuals only undertake survey work that is within their area of competence.

2.5   The professional competence of ecologists to be engaged on a project may be gauged by their attainment of a combination of credentials including one or more of the following:

- relevant academic qualifications;
- proven track record in ecological field surveys and/or specialist taxonomic fields;
- membership of an appropriate professional body;
- fulfilment of continuing professional development requirements;
- internal training programmes and informal examination schemes; and
- supplementary qualifications, for example, identification qualifications.

2.6   Details of entry requirements, code of ethics and continuing professional development requirements can be obtained from the professional institutes listed in Appendix 6.

Consultation and Data Collection

2.7   Appendix 2 lists the statutory consultees who should be contacted at the beginning of the scoping process to help define the likely significant ecological impacts of the proposed development and also to identify existing data which could assist in defining the baseline ecological conditions.

2.8   Appendix 2 also lists some of the other organisations which may hold relevant ecological data and may be able to assist in defining the scope of the ecological study. Donn and Wade (1994) have published a comprehensive county by county list of ecological information sources and their contact addresses. In addition to these sources, previous experience has shown that local naturalists can frequently provide some of the most extensive and reliable data.

However, it should be appreciated that the time and resources of many local specialists and many of the organisations listed by Donn and Wade (1994) are often severely limited and may easily be overwhelmed by constant requests for information and advice. Access to certain information, such as the location of rare and protected species, may also be restricted and charges for information should be expected.

2.9   The desk study should also encompass published sources of information such as the Ancient Woodland Inventory (Spencer and Kirby, 1992) and county and national atlases of flora and fauna (eg Harding, 1993) and the Invertebrate Site Register (ISR) compiled by the JNCC.

2.10   When undertaking a desk study and reporting the results it should be recognised that the quality of ecological data from different sources is highly variable (Wyatt, 1991a; 1991b). Typical criticisms include the lack of information on when a survey was undertaken, the type of survey method(s) used (if any) and the overall competence of the ecologists carrying out the work. Where such gaps in information exist, or the quality of data is suspect for other reasons, an on-site survey may be necessary to update and/or verify the coverage and accuracy of previous surveys and collected data. Within an Environmental Statement the survey date and reliability of any existing information used in the assessment should also be clearly stated.

Both badgers and their setts are protected by law. Despite this they are threatened by digging and baiting in many areas, and sett locations should remain confidential

TABLE 1 STATUTORY AND NON-STATUTORY SITE DESIGNATIONS

International Obligations	World Heritage Site	Convention Concerning the Protection of the World's Cultural and Natural Heritage 1972
	Biosphere Reserve	Man and Biosphere Programme (MAB) UNESCO
	Special Protection Area (SPA)	EC Wild Birds Directive 79/409
	Ramsar Site	Convention on Wetlands of International Importance especially as Waterfowl Habitat 1971 (Ramsar Convention)
	Special Area of Conservation (SAC)	EC Directive on the Conservation of Natural Habitats and of Wild Fauna and Flora (92/43/EEC) (EC Habitats Directive)
	Biogenetic Reserve	Council of Europe Programme for a European Network of Biogenetic Reserves
National Statutory Designations	National Natural Reserve (NNR)	Wildlife and Countryside Act 1981
	Marine Nature Reserve (MNR)	Wildlife and Countryside Act 1981
	Area of Special Protection for Birds (AOSP)	Wildlife and Countryside Act 1981
	Sites of Special Scientific Interest (SSSI)	Wildlife and Countryside Act 1981: Nature Conservation and Amenity Lands (Northern Ireland)Order 1985
	Areas of Special Scientific Interest (ASSI)	Nature Conservation and Amenity Lands (Northern Ireland) Order 1985
	Environmentally Sensitive Area (ESA)	Agricultural Act 1986

	Limestone Pavement Order	Wildlife and Countryside Act 1981
	Nature Conservation Order	Wildlife and Countryside Act 1981
	Marine Consultation Area	
Local Site Designations	Local Nature Reserve (LNR)	National Parks and Access to the Countryside Act 1949
	Local Authority or Wildlife Trust Non-Statutory Site (Site of Bird Interest: Site of Nature Conservation Importance: Site of Importance for Nature Conservation)	
Nature Reserves	County Wildlife Trust Local Authority Royal Society for the Protection of Birds Woodland Trust National Trust	
Other	National Park	National Parks and Access to the Countryside Act 1949
	National Scenic Area	Town and Country Planning (Scotland) Act 1972
	Natural Heritage Area (NHA)	National Heritage (Scotland) Act 1991
	Area of Outstanding Natural Beauty (AONB)	National Parks and Access to the Countryside Act 1949; Amenity and Landscape Act 1965 (Northern Ireland)
	Heritage Coast Local Landscape Designation Ancient Woodland Site	

EXTENDED PHASE 1 HABITAT MAP WITH TARGET NOTES

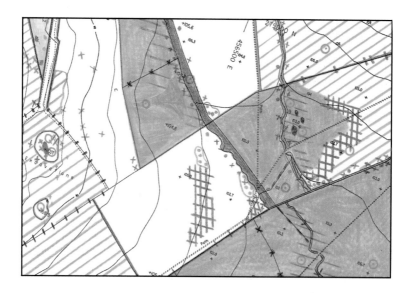

EXAMPLE TARGET NOTES

8   Strip of broadleaved woodland with beech, sycamore and oak over scrub of hazel and hawthorn and a ground flora including wood sorrel, male fern and dogs mercury. Several mature oak and beech have woodpecker holes, loose bark and occasional hollow branches: hence potential for bats and hole nesting birds. Numerous piles of dry deadwood where ground and stag beetles were found. Singing nuthatch and yellowhammer present.

2   Marginal vegetation includes tufted hair grass, soft rush, ragged robin and great willow herb with scattered scrub of hawthorn dragonflies, hoverflies, bush crickets and stoneflies present.

10   Disused quarry with a small shallow-sided pond and scattered scrub including willow, hawthorn, bramble and gorse. The pond has marginal vegetation including soft rush, great willow-herb; young newts also seen, and tracks in the mud included deer (muntjac) and moorhen. The quarry sides are rich in bryophytes and lichens and have ferns including lemon-scented fern, hard fern and lady fern.

Example of a Phase 1 map (JNCC 1993) with Extended Target Notes

2.11   Consultations with the general public are also beneficial for identifying issues of local concern, and can on occasions reveal different issues from those initially identified by technical experts.

2.12   During the above consultations such issues as client confidentiality and limited access to certain information (eg the location of badger setts) should be taken into account.

2.13   Recognised sites of nature conservation interest (statutory and non-statutory) which are within or near to areas likely to be directly or indirectly impacted, should be identified within an Environmental Statement and the main reasons for their designation outlined. A minimum area of search of a 2km radius around the development site is usually appropriate for obtaining information, although this may need to be extended where the impacts may be over a much larger area (eg air quality impacts of power stations).

2.14   The main types of site designations that may need to be considered within a baseline description are listed in Table 1. The list includes statutory and non-statutory designations since many planning authorities now have policies aimed at protecting local sites of conservation interest as well as nationally important statutory sites and protected species. Furthermore, it is more likely that non-statutory sites for nature conservation will be encountered in an EA than higher tier sites, such as NNRs and SSSIs, which ideally have been excluded at the site selection stage of an EA. Collis and Tyldesley (1993) provide a useful review of the existing status, policy framework and evaluation criteria for non-statutory Sites of Importance for Nature Conservation (SINC).

2.15   Whilst sites of international and national importance are usually designated according to strict selection criteria, in certain cases sites of local conservation importance have been subject to less rigorous scientific and objective appraisal. To verify the conservation status of designated sites, accompanying information should therefore be requested on:

- •   the criteria used in evaluating the site;
- •   the date of designation;
- •   the date and results of any subsequent surveys or modifications to a site's status.

2.16   Although the identification of designated sites of conservation interest is important for evaluating the baseline environment, care should be taken that an ecological assessment does not place undue emphasis on the presence of these sites at the expense of wider interests. This is because wildlife conservation is reliant upon the protection of the wider countryside in conjunction with a system of individual site designations.

Site Visit and Extended Phase 1 Survey

2.17   Maps showing the habitat structure of the whole site likely to be directly impacted by the development should be presented with habitats classified according to the Phase 1 habitat methodology (JNCC, 1993). In some cases recent Phase 1 survey information already exists and can be utilised as a basis for the survey.

2.18   The maps should be accompanied by target notes which identify and provide further information on habitat features of particular value to different ecological groups such as plants, fungi, lichens, mosses and birds.

2.19   The target notes for such an extended Phase 1 habitat survey should include relevant information on features of conservation interest for each ecological group (see illustration, pg.14). This combined approach is necessary as the habitat requirements of different groups and species of flora and fauna are highly variable and not always deducible from a purely botanical habitat survey. For example, habitats supporting species-rich invertebrate communities can be of low botanical interest.

2.20   Even though simple Phase 1 habitat information for the site may exist on file, it is unlikely that these data will contain the necessary target notes highlighting the value of the various habitats for different faunal and floral communities. In most cases, therefore, a site visit will be necessary to gather the required information.

2.21   The geographical extent of the area to be covered by this assessment will vary enormously depending on the type of project and location of the development. It is important to gather enough information on the initial visit (or from a desk study) to be able to set the development site into context with the surrounding area. Moreover there needs to be sufficient information presented to identify important communities outside the boundary. For example, noise disturbance/hydrological impacts might affect important breeding bird communities some distance from the site boundary, or new road schemes might intersect badger feeding areas. The existence of these communities adjacent to the development site needs to be determined as part of the initial assessment.

2.22   Where habitats forming part of a river corridor are likely to be directly or indirectly impacted, it may also be appropriate to present data according to the NRA river corridor survey method (NRA, 1992). This is based on mapping the major habitats, vegetation and physical features of 500m linear sections of river corridor. However, the method is principally directed at mapping habitats with notes on dominant plant species and as a consequence limited information is provided on other ecological groups. It is imperative, therefore, that target notes, as described above, are included as part of the survey.

2.23   Baseline surveys for ecological assessments in marine and estuarine habitats often need to be more extensive than those in terrestrial habitats because of the importance of indirect impacts arising from changes in coastal processes. The extent of the baseline survey area should, therefore, be determined where possible by the predicted changes to the physical environment, including factors such as tidal flows, sediments and water quality. In these habitats, the information which can be collected at Phase 1 level may be limited by practical considerations.

2.24   For subtidal habitats on hard coasts, the Phase 1 habitat survey method (JNCC, 1993) is not appropriate and the SEASEARCH method (Earll, 1992) should be used where possible. This is a visual appraisal method using divers and provides a description of the seabed topography and habitats present. It can only be used in sites where it is safe or appropriate to use divers. Sites and habitats are categorised using the Marine Nature Conservation Review (MNCR) classification.

2.25   There is no equivalent Phase 1 survey method for soft subtidal sediments and it is, therefore, necessary to use information from consultation and desk study.

2.26   For intertidal and coastal fringe habitats the Phase 1 habitat survey method (JNCC, 1993) may be used. However, in some cases, it may be more appropriate to map the habitat using the MNCR classification or Coastwatch survey method (Hiscock, 1990).

Production of Ecological Scope

2.27   Given the above information it should then be possible to determine whether the proposed development will have any significant ecological impacts. If significant impacts are anticipated then using the information gathered it should be possible to scope the extent of the ecological baseline information required (ie not available from existing records) in order to fully assess the impacts.

2.28   In some countries (eg Netherlands, Indonesia and Canada) independent scoping of an EA is carried out by a government agency. No such government organisation exists in the UK, but advice on scoping can be obtained from the IEA and the statutory consultees listed in Appendix 2.

2.29   At the end of the scoping process it is good practice to summarise in a short report the proposed scope of the ecological assessment and circulate it to all the statutory consultees and other key organisations identified during the consultation process. Obtaining agreement by these organisations that the scope of the proposed study is adequate is an important stage in the assessment

Figure 1: Steps in the process of baseline ecological assessment

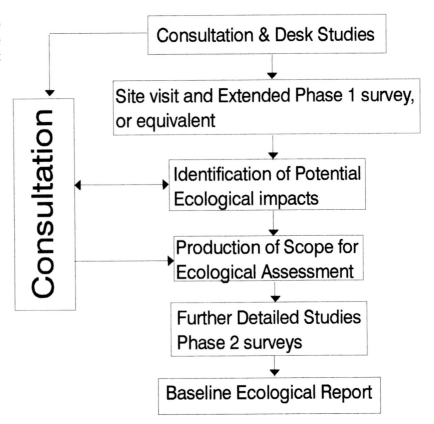

process. If this is not achieved then objections as to the adequacy of the Environmental Statement can be expected at a later stage when it is submitted for consideration to the Planning Authority. Resolving such objections can result in substantial delays in determination of planning approval or rejection (Coles and Smith, 1993).

Summary

- A scoping exercise to focus effort onto the most important elements of the EA is essential. Scoping the ecological elements of these studies is, however, an iterative process and involves some initial work in order to determine whether the development will significantly impact any important ecological features.

- Determination of the potential ecological impacts of a development requires a desk study involving consultation and data gathering, and the preparation of maps showing the habitats of the impacted site with target notes identifying the main features of conservation interest for plants, lichens, fungi, mosses, birds, invertebrates, amphibians, reptiles and mammals namely an extended Phase 1 survey.

- The extent of the area covered by this initial site visit should be sufficient to evaluate the importance of the site in its local context and to identify any communities affected by the development beyond the site boundary.

- The NRA river corridor methodology together with extensive target notes as for the extended Phase 1 survey is proposed as the method for presenting Phase 1 information relating to rivers.

- SEASEARCH method using divers can be used, where practical, for some subtidal habitats. The Marine Nature Conservation Review provides suitable categories for habitat mapping in marine and estuarine environments.

- The information collected in this exercise should enable an ecological scope to be produced. Agreement on the scope with statutory consultees can prevent objections as to its adequacy at a later stage.

General Principles

Background

3.1   The previous Section focused on identifying the minimum level of information that should be included within the ecological baseline. One of the main purposes of these preliminary studies is to determine the need for further, more detailed surveys (Phase 2 surveys) of particular ecological groups.

3.2   Phase 2 surveys typically involve a far greater input of time and resources as well as the use of relevant specialists. Nevertheless where there are no existing data and additional information is required because of the likely impact significance, then Phase 2 studies will be necessary.

3.3   The following Sections therefore describe:

•   criteria or situations where more detailed surveys should normally be undertaken;
•   survey methods for different ecological groups;
•   factors for consideration in the evaluation of baseline data.

3.4   In this Section a general introduction is provided on the use and scope of the proposed survey criteria, the key factors that should be taken into account when planning and conducting ecological surveys and the different approaches used to evaluate baseline data.

3.5   For ease of reference the guidelines describe in separate sections the survey criteria and methods for different ecological groups comprising vegetation, birds, mammals, amphibians and reptiles, fish and terrestrial and aquatic invertebrates. A separate section also covers marine and coastal habitats and organisms since the scope of the required surveys and in some cases the methods used are different from those of terrestrial environments.

3.6   Although the guidelines consider separately the various ecological groups, an ecological assessment should always adopt an *ecosystem perspective* and highlight any key relationships that exist between different species and the surrounding environment. Critical factors may include, for example, soil series, hydrology, topography (aspect and slope), microclimate and management regimes (eg grazing, seasonal mowing, coppicing), for terrestrial systems, or water quality and flow profiles for aquatic systems.

Survey Criteria

3.7   For each ecological group, criteria are outlined to indicate the situations when more detailed surveys will normally be required. The proposed criteria do not attempt to address specific impact situations but rather to identify general considerations. For any particular project the criteria should therefore be interpreted with the project specification in mind.

3.8   Many of the criteria used include reference to protected species and habitats. Appendix 3 provides a summary of the species of flora and fauna which are protected in the UK under international obligations and national and European legislation.

3.9   Throughout the text references to the Wildlife and Countryside Act and Wildlife Order NI also include subsequent amendments and quinquennial reviews. For all legislative requirements it is imperative that the most up-to-date acts, schedules and amendments are referred to.

3.10   Information on national and local planning policies for nature conservation is given in:

- Government Circulars (eg Department of the Environment (DoE), 1987; 1992a).
- Planning Policy Guidance Notes (eg DoE, 1988; 1992b; 1994).
- Policy statements such as the UK Action Plan on Biodiversity (DoE, 1994) and the UK Strategy for Sustainable Development (DoE, 1994).
- General texts such as Countryside Commission (1990), Countryside Commission et al. (1993), DoE (1993) and Collis and Tyldesley (1993).

3.11   The recommended survey criteria also refer to Red Data Book (RDB) species. These are nationally rare species and are allocated a threat status according to the World Conservation Union (IUCN) RDB categories.

Viola persicifolia

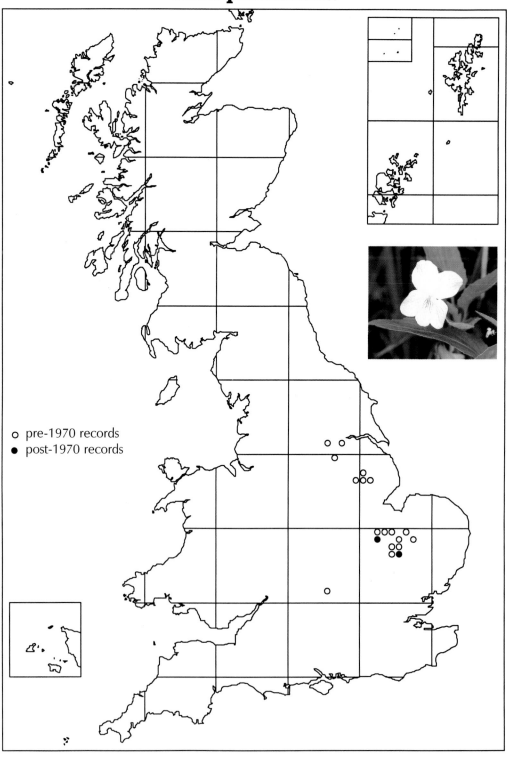

○ pre-1970 records
● post-1970 records

RDB species distri-
bution can be
plotted on maps using
10km grid squares

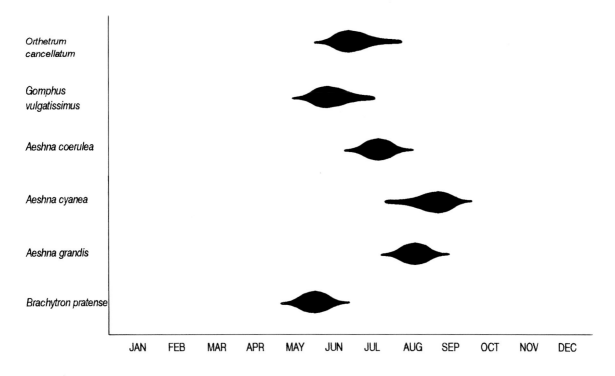

Figure2: A Gant diagram showing flight periods for various species of dragonfly

Survey Methods

3.12 Recommended survey methods should be used for each ecological group to ensure that the data collected during any proposed surveys are comprehensive and can be easily verified. For most ecological groups, the time of year when a survey is undertaken can also significantly affect the quality of the collected data in terms of its coverage, level of detail and accuracy. The importance of carrying out field surveys for different species at an appropriate time of year is illustrated in Figure 2.

3.13 In all Environmental Statements survey date(s) should be clearly stated. For situations where a survey has been conducted at an inappropriate or suboptimal time of year the reasons for this should be presented in the Environmental Statement together with an indication of the reliability of the collected data.

3.14 All field surveyors must ensure that they remain within the law relevant to the species they are studying and that they do not cause unnecessary stress to animals during a survey. Details of the licensing procedures for the study of protected species are given in Appendix 4. It should be remembered that all land is owned by somebody and that permission to carry out wildlife surveys will need to be obtained from the owners.

Evaluation

3.15   Impact significance is a product of the magnitude and scale of an impact, the assessed importance of the species or habitat(s) likely to be affected and its/their sensitivity to the predicted impact (English Nature, 1994). It is therefore important that an ecological baseline provides sufficient information to evaluate the ecological and nature conservation importance of a survey area and/or species of concern. Although criteria and methods for the evaluation of nature conservation importance have been discussed at length and are well documented (Helliwell, 1973; Ratcliffe, 1977; Margules and Usher, 1981; Usher, 1986; NCC, 1989; Collis and Tyldesley, 1993), evaluating nature conservation interest remains a difficult and complex process not least because of the inherent inclusion of value judgements and subjective considerations. In response to these difficulties, Sections 4-10 of the guidelines outline general features that should be considered in the evaluation of baseline data collected for each ecological group.

3.16   For species, baseline evaluation criteria may include reference to protected species (See Appendix 3), rare species or those species that are important in the functioning of an ecosystem (keystone species). With regard to rarity, this is usually considered at more than one geographical scale, eg national, regional and local (the latter usually taken to mean vice-county or district). At the national scale lists of nationally rare (RDB) and scarce species are available, whilst at the county level, details of locally rare and scarce species can often be obtained from: the lead body responsible for non-statutory site designation; from local biological record centres; from the relevant county recorder; or from publications such as county floras, avifaunas and atlases.

3.17   For sites, the scientific evaluation criteria described by Ratcliffe (1977) and developed further in the guidelines for the selection of biological SSSIs (NCC, 1989) may be applied. These selection criteria are principally directed at identifying and evaluating statutory sites of national importance. Guidance on the criteria used in the evaluation of non-statutory sites is given in Collis and Tyldesley (1989). The latter have shown that non-statutory site systems are mostly county based and represent a partnership usually between planning authorities, the relevant county wildlife trust or urban wildlife group and the local English Nature office (or country equivalent). Many areas, such as the West Midlands, Derbyshire and Avon, have published registers of non-statutory sites and/or can make available the criteria which have been used to select designated sites.

3.18   For both statutory and non-statutory sites, an important consideration in the evaluation of baseline data is the presence of undisturbed ancient semi-natural habitat (eg ancient woodland).

An ancient semi-natural woodland in lowland Eastern England, showing a typical suit of species (Bradfield)

3.19   Of particular relevance and importance in the evaluation of non-statutory sites are social or community factors. The consideration of community factors was originally associated with the evaluation of urban wildlife habitats (eg locations in areas of ecological deficiency, wildlife corridors, networks), but now includes criteria that relate to social and amenity uses (eg education, private study, amenity) and is increasingly relevant to rural situations. Generally, there are few objective measures for community criteria, but possible evaluation approaches could include numbers of people visiting a site, the frequency of visiting school parties, number of access footpaths, or opinion surveys.

3.20 In a wider context, the evaluation of baseline data should also address the increasing emphasis placed on sustainable development (English Nature, 1992; DoE, 1994). As an example, this could mean maintaining the environment's natural qualities and characteristics and its capacity to fulfil its full range of functions, including the conservation of biodiversity. Those aspects of biodiversity which cannot readily be replaced are considered to be critical natural capital. These include ancient semi-natural habitats that depend for their survival upon traditional kinds of land management (eg ancient coppice woodlands, hay meadows, and chalk/limestone grassland). The principles of sustainable development require that critical natural capital, comprising both statutory and non-statutory sites, must be protected. Other aspects of biodiversity which should not be allowed, in total, to fall below minimum levels, but which could be created elsewhere within the same area, are referred to as constant natural assets.

Summary

- The requirement for Phase 2 ecological surveys should be determined with reference to preliminary ecological studies and recommended survey criteria for each ecological group.

- An ecological assessment should adopt an ecosystem perspective and highlight any key relationships that exist between different species and the surrounding environment.

- For protected species and habitats, reference should be made to the most recent international and national legislative requirements.

- Ecological surveys should follow appropriate survey methods and should be carried out at the correct time of year by competent ecologists to provide comprehensive and reliable survey data.

- Field surveyors must ensure they comply with the relevant licensing procedures for studying protected species.

- The significance of an impact is a function of the magnitude and scale of the impact, the assessed importance of the species or habitat(s) likely to be affected and its/their sensitivity to the predicted impact.

- An ecological baseline should provide the necessary information with which to evaluate the ecological and/or nature conservation importance of the species and/or area likely to be impacted.

- Species evaluation criteria may relate to protected, rare or keystone species; rarity should be considered at national, regional and local scales.

- Site evaluation may include reference to scientific evaluation criteria (Ratcliffe, 1977; NCC, 1989), the presence of ancient/semi-natural habitats and the social/amenity value of a site to the local community.

- Baseline ecological input should enable an evaluation of a site to be made in the context of environmental sustainability, for example, by identifying critical natural capital and constant natural assets.

PART THREE

Detailed criteria, survey
methods and evaluation

Vegetation

Introduction

4.1   The term vegetation encompasses a wide range of taxa including flowering plants and ferns (vascular plants); lichens; mosses and liverworts (bryophytes); and algae including stoneworts. In this Section the criteria and methods for Phase 2 surveys are described for vascular plants, lichens, bryophytes and freshwater algae since these groups are most likely to require survey in their own right. It should be emphasised that although each group is addressed in these guidelines, for most ecological assessments only Phase 2 surveys of vascular plants are usually required. Detailed surveys of lichens and algae have seldom been carried out and tend to be confined to a limited number of circumstances. For example, lichens may be used to monitor environmental conditions due to their extreme sensitivity to atmospheric pollution; freshwater algae are similarly sensitive to nutrient enrichment and pollution of water bodies, so may be used to monitor water quality. Fungi are now recognised as a separate kingdom but are usually recorded in conjunction with surveys of vascular and non-vascular plants. Phase 2 surveys principally concerned with fungi are very rare. It has not been possible to include a section on fungi in these guidelines, (see Introduction).

Vascular Plants

Survey Criteria for Vascular Plants

4.2   More detailed surveys of vascular plants should be conducted under the following circumstances:

*   When the desk study or extended Phase 1 survey indicates that a development may affect any plant species listed in the Red Data Book (RDB) (Perring and Farrell, 1983; Curtis and McGough, 1988); Schedule 8 of the Wildlife and Countryside Act (Wildlife Order NI); Appendix I of the Bern Convention; Annexes II and IV of the EC Habitats Directive (see Appendix 3); or a Nationally Scarce plant species, as listed in Stewart *et al.* (in press).

*   When the desk study or extended Phase 1 survey identifies habitats of statutory significance for vascular plants. These

include, for example, habitats of community interest and especially priority habitats (Table 2) as listed in Annex I of the EC Habitats Directive (Appendix 5).

- When the desk study or extended Phase 1 survey identifies vegetation types of potential regional or local importance.

4.3   The inclusion of species in the British RDB (Perring and Farrell, 1983) is based on the occurrence of native species recorded in 15 or fewer 10km grid squares covering Great Britain. Nationally Scarce species are defined as those occurring in 16-100 10km grid squares in Great Britain. For Ireland, RDB species are described in Curtis and McGough (1988).

Old parklands and trees can often serve as important habitats for mosses and lichens as in this example

TABLE 2 PRIORITY EC HABITATS IN THE UK*

- coastal lagoons
- inland salt meadows with *Puccinellia distans*
- fixed (grey) dunes
- eu-Atlantic decalcified fixed dunes with *Calluna vulgaris*
- decalcified fixed dunes with *Empetrum nigrum*
- dune Juniper thickets
- dry coastal heaths with *Erica vagans*
- wet heaths with *Erica ciliaris* and *Erica tetralix*
- dry calcareous grasslands (*Festuco-Brometalia*) important for orchids
- species-rich *Nardus* grasslands
- active raised bogs
- active blanket bogs
- calcareous fens typified by *Cladium mariscus*
- tufa springs with moss vegetation (*Cratoneuron* species)
- alpine pioneer formations with *Carex atrofusca*
- limestone pavements
- *Tilia cordata - Acer campestre* ravine woodlands
- Caledonian pine forest
- bog woodland (natural types only)
- *Taxus baccata* woods
- residual alluvial woods (*Alnus glutinosa - Prunus padus* type)
- Mediterranean temporary pools

* Further information see Appendix 5.

Survey Methods for Vascular Plants

4.4   Further surveys of vascular plants should be carried out and the results presented in accordance with the National Vegetation Classification (NVC) (Rodwell, 1991; 1992a; 1992b; in press (a); in press (b)). The NVC provides a standardised and systematic means of recording and classifying the vegetation of all natural, semi-natural and major artificial habitats within the UK. The survey method involves detailed species recording (including vascular plants, bryophytes and macro-lichens) using quadrats of varying sizes and assigning a measure of percentage cover to each species recorded within the quadrats. In conjunction with the recorded quadrat data, an additional species list for the entire study site is normally presented. From this information the NVC plant communities and sub-communities described in the various handbooks can be determined. Software packages, such as MATCH (Malloch, 1991) and TABLEFIT (Hill, 1993), can be used to ascribe NVC community types to a given species list and indicate the accuracy of the fit, although reference should always be made to the published text before any final decision is reached.

4.5   Although recommended for most situations, there are certain habitats where the NVC is difficult to apply or inappropriate. For example, it can be difficult to determine the relevant NVC plant communities in transitional and disturbed habitats. For certain habitats there are also specific survey methods which may be more suitable depending on the aims of the survey. These survey methods include: Alcock and Palmer (1985) for ditches, dykes and rhynes; Palmer *et al.* (1992) for standing waters and canals; Holmes (1987) for rivers; Clements and Tofts (1992) for hedgerows; and Kirby (1988) for woodlands. Survey methods for marine and coastal vegetation are described in Section 10.

4.6   For sites where the presence of protected or rare species is the main issue of interest, a detailed survey of the species distribution and abundance would be more appropriate than undertaking a general survey of the plant communities present.

Lichens and Bryophytes

Survey Criteria for Lichens and Bryophytes

4.7   More detailed surveys of lichen and bryophyte communities should be conducted under the following circumstances:

- When the desk study or extended Phase 1 survey indicates that a development may affect species listed in the Red Data Book (RDB) for Britain and Ireland (Stewart and Church, in press (a); in prep (b)), an assemblage of Nationally Scarce species (Hodgetts, 1992), or those listed in Schedule 8 of the Wildlife and Countryside Act (see Appendix 3).

- When the desk study or extended Phase 1 survey identifies habitats of particular importance for lichens or bryophytes and the development may cause a significant impact on the communities either through direct loss or indirect impacts such as deterioration in the air quality or nutrient enrichment. Potentially important habitats for these groups include old woodlands/parklands, mine spoil heaps, oceanic sites in the west, north-western Highland hills, sand dunes, fens, bogs, heathland, chalk downland, river gorges and flushes.

4.8   The inclusion of species in the RDB has been based on the occurrence of native species recorded in 15 or fewer 10km grid squares in Great Britain or in 10 or fewer squares in Ireland. Nationally Scarce species of lichens and bryophytes are defined as those recorded in 16-100 10km grid squares in Great Britain.

Survey Methods for Lichens and Bryophytes

4.9   No standard quantitative technique exists for surveying lichens or bryophytes. However, semi-quantitative studies should be carried out where possible and for lichen communities should be described with reference to James *et al.* (1977) and to data held on the distributional atlases published by Hill *et al.* (1991). Where important lichen communities are identified, a photographic baseline monitoring programme should be implemented (Hawksworth and Rose, 1976; Richardson, 1992).

Freshwater Algae

Survey Criteria for Freshwater Algae

4.10   More detailed surveys of aquatic algae should be conducted under the following circumstances:

- When a predictable pollution event or development is likely to affect a species listed in the Red Data Book (RDB) on stoneworts (Stewart and Church, 1992), a Nationally Scarce species (Hodgetts, 1992) or those listed in Schedule 8 of the Wildlife and Countryside Act (Wildlife Order NI) (see Appendix3).

- When the development will cause gross changes to the algal community such that these changes will have a large impact on other groups within the ecosystem.

4.11 The only RDB on British and Irish algae covers the stoneworts (Stewart and Church, 1992), with species selection fulfiling at least one of the following criteria: species found in Great Britain in 15 (10 in Ireland) or fewer 10km grid squares since 1970; species with slightly more records than this, but whose populations are known to be low at all or nearly all sites; species with more records than this but which have shown a marked decline; species that are still under-recorded, but are known to be sufficiently habitat confined that they are likely to fit one or more of the other criteria. Nationally Scarce species are those recorded in 16-100 10km grid squares in Great Britain since 1950.

Survey Methods for Freshwater Algae

4.12 No standard method for undertaking general surveys of macro-algae including stoneworts exists though one for assessing water quality using aquatic macrophytes (including macroalgae) has been proposed by Holmes (1987).

Evaluation

4.13 Where populations of protected, RDB or Nationally Scarce vascular or non-vascular plants may be impacted, the number of sites supporting a population and the size of each population, both in a local and national context, should be determined. This will provide a measure against which to compare the population size/number of sites likely to be impacted. At a national scale sources of distributional data include: Perring and Walters (1962), Perring and Sell (1968), Graham and Primavesi (1993) and Preston *et al.* (1993) for vascular plants; Hill *et al.* (1991; 1992; 1994) for bryophytes; Seward and Hitch (1982) for lichens; and Moore and Green (1983), Moore (1986) and Stewart and Church (1992) for charophytes (including stoneworts). To provide a local context for evaluation, reference should be made to the appropriate county flora and/or the local BSBI recorder.

4.14 An additional means of evaluating a site supporting RDB or Nationally Scarce vascular or non-vascular plants is to apply the criteria for determining whether a site achieves SSSI status in terms of species interest and is therefore of national importance. The evaluation method uses scores for species, which are summed and compared to a SSSI threshold (NCC, 1989; Hodgetts, 1992).

4.15 Evaluating the conservation importance of habitats is complicated by the lack of accurate and comprehensive data on the extent and distribution of all habitats and vegetation types, except for certain well surveyed and researched examples. Estimates of the extent of different plant communities are given in NCC (1989) and some information on their distribution and rarity are given in Rodwell (1991; 1992a; 1992b; in press (a); in press (b)).

Summary

- Further surveys of vascular plants should be undertaken where protected, RDB and/or Nationally Scarce species may be affected, or where habitats of national or international nature conservation interest or other semi-natural habitats, particularly those of regional or local importance, lie within the impact area.

- Further surveys of lichens and bryophytes should be undertaken where protected, RDB or Nationally Scarce species may be affected, or where habitats recognised as having significant conservation interest for lichens and bryophytes lie within the impact area and the development may cause significant impact on the communities.

- Further surveys of freshwater algae should be undertaken where there is a predictable impact on a protected, RDB or Nationally Scarce species of stonewort.

- Where protected, RDB or Nationally Scarce species may be impacted, the population size/number of sites likely to be impacted should be compared to national and local baselines.

- Sites containing such species may be evaluated using the criteria to determine whether SSSI status is achieved in terms of species interest.

Birds

Survey Criteria

5.1 More detailed bird data should be presented under the following circumstances:

- When the desk study indicates that a breeding pair or population of a species which occurs on Schedule 1 of the Wildlife and Countryside Act (Wildlife Order NI) or Annex 1 of the EC Birds Directive (see Appendix 3) may be affected by the development.

- When the extended Phase 1 survey indicates that suitable breeding habitat is available on the development site in proximity to known populations of Schedule 1 or Annex 1 species. For example, where reedbeds for which no information is available occur in regions known to have breeding bearded tit, marsh harrier or bittern, or mature pine plantations occur in a county or region with breeding crossbill populations.

- When the desk study indicates that the development may have a significant impact on an area of continuous habitat, for example salt marsh, or of discontinuous but linked habitat, for example a group of flooded gravel pits. The development may be of potential significance for species of nature conservation interest, or lie within the boundaries of a site known to contain 1% or more of any species in the UK or a 'nationally important' number where other criteria apply. Thus where a marina development is proposed on an estuary which holds more than 1% of the relevant passage or wintering populations of one or more waterfowl species, or a mean minimum wintering total of at least 10 000 waterfowl, the effect of direct or indirect loss of feeding grounds or roosting areas on these birds should be considered. Data on the roosting, feeding and loafing areas for waterfowl both on the development site and throughout the estuary would need to be presented in order to evaluate the importance of the site.

- When the desk study or extended Phase 1 survey identifies within the impact area vulnerable habitats that are associated with a particular assemblage of scarce breeding birds. An example of such an assemblage would be wet meadows holding breeding lapwing, snipe, redshank and curlew. Vulnerable habitats would also include:

(i)   montane areas above 600m
(ii)   heathland and brecks
(iii)   bogs, moors, swamps and fens
(iv)   salt marsh, intertidal mudflats, dunes, shingle and sea cliffs
(v)   native pine woods
(vi)   lowland wet grassland
(vii)   machair

•   Where the desk study or the extended Phase 1 survey indicates that the area to be impacted by the proposed development contains species or populations important at a local or regional level.

5.2   For many sites and species these data may already exist and additional surveys will therefore not be required. Where existing data are insufficient, additional surveys should be targeted on the population/season of interest.

Survey Methods

5.3   Where a development is likely to affect scarce breeding species appropriate survey techniques and timing will depend on the species and habitat concerned. Bibby *et al.* (1992) describe appropriate methods for a wide range of species. Specific methods are available for seabird breeding colonies (Lloyd *et al.*, 1991).

5.4   Where the development is likely to affect an assemblage of local or regional importance, most survey methods will typically involve identifying breeding behaviour as defined for atlas studies (eg Gibbons *et al.*, 1993) whilst walking over the site on one or more suitably timed visits.

5.5   Use of the full Common Bird Census (British Trust for Ornithology, 1983) or Waterways Birds Survey (British Trust for Ornithology, 1982) methods, which involve eight or more visits during the breeding season, are only of value when the location of breeding territories for all species present on a site is important. Neither method, however, adequately surveys most scarce species (with the exception of certain passerines, such as the Dartford

40

warbler) and they are rarely cost-effective in the ecological assessment context.

5.6　If the development is likely to affect roosting or feeding areas used for wintering or on migration then data from existing records wherever possible should be presented for the peak usage of the impacted areas over the last 5 years. If no such data exist then monthly count data, covering the seasons of importance for birds in the impacted area, should be collected. Specific methods are available for estuarine habitats (high water roosts: British Trust for Ornithology, undated (b); low water counts: British Trust for Ornithology, 1992) and for counting seabirds at sea (Tasker *et al.*, 1984).

Oystercatchers coming
in to roost

Evaluation

5.7　Where breeding populations may be impacted the size of each population, as a percentage of the local/regional populations and the UK population, and the distribution of nest sites/breeding territories/home ranges in relation to the total suitable habitat available should be determined where possible. Estimates of British and Irish populations are given by Gibbons *et al.* (1993) and estimates of county or district populations may be derived from atlases, avifaunas and local bird reports.

5.8　Where wintering populations or birds on migration may be impacted the size of each population, as a percentage of the local/regional populations and the relevant UK/world population, should be determined. Estimates of UK wintering populations and relevant world populations for waterfowl are given in annual Wildfowl and Wader Count Reports (published by the Wildfowl and Wetlands Trust) and estimates of county or district populations by local bird reports. Estimates of seabird populations away from the breeding sites are given in Tasker *et al,* (1987), Webb *et al.* (1990) and Stone *et al.* (in press).

Summary

- Further surveys of birds should be undertaken where protected species may be affected, or where a habitat holding at least 1% of the UK population of a species (or other recognised nationally important number) may be impacted, or where a vulnerable habitat associated with a particular assemblage of breeding birds lies within the impact area, or for sites where there are species or populations of importance at a local or regional level.

- Breeding bird surveys, including their timing, must be targeted on the species or habitat of interest.

- Wintering bird surveys should present if possible several years' data from existing records; where not available monthly counts identifying both roosting and feeding sites should be undertaken.

- Breeding bird population data should be presented as a percentage of both UK and local/regional totals (where available) at the appropriate season.

- For wintering populations or birds on migration, baseline data should be presented as a percentage of the local/regional populations and the UK/relevant international population.

Mammals

Survey Criteria

6.1 More detailed information on mammals should be presented under the following circumstances:

- Where the desk study or extended Phase 1 survey indicates the probable presence of species protected under the Bern, Bonn and ASCOBANS conventions, the Wildlife and Countryside Act (Wildlife Order NI), the EC Habitats Directive (see Appendix 3) or the Protection of Badgers Act 1992 (this Act consolidates British legislation on the protection of badgers and their setts).

- Where mammals are likely to interact with the operation of a development. An example could be the potential hazard caused by deer and badgers crossing new roads.

- Where a population of mammals has an important function within the ecosystems in and around the proposed development, and where the impact of the development will be so great as to alter the population of mammals to such an extent that there are significant ecosystem changes.

- In cases where consultations with interested bodies identify a reasonably founded local perception of the importance of species not otherwise protected through legislation.

Survey Methods

6.2 Survey methods for mammals will be different for different species and will also vary as a result of the specific requirements of individual studies. The descriptions of the methods given below are intended to give a broad indication of the different types of survey methods available. For surveys of many mammal species, the surveyor needs to be licensed (see Appendix 4).

6.3 **Badger**. Survey methods for badgers are given in Cresswell *et al.* (1990), Harris *et al.* (1989) and Kruuk (1978). Of particular relevance to the Protection of Badgers Act 1992 and the Wildlife (NI) Order is the presence of setts. The most effective time to undertake surveys of setts is during the winter when vegetation is less dense. Surveys of feeding areas and habitual runs are best carried out in the spring, summer and autumn months, when the animals are more active.

6.4 **Bats.** Bat surveys, as outlined by NCC (1987), should identify roost and hibernation sites. Evidence of feeding areas and important flight routes may also be required in some studies. The most appropriate time to survey for roosting sites and feeding areas is April to October. Hibernation sites are best surveyed from November to March.

6.5 **Dormouse.** Details of methods for conducting dormouse surveys are given in Bright *et al.* (in press) and Bright and Morris (1990). Such surveys are best conducted through September to December. Population studies require the use of nestboxes and should be undertaken outside the hibernation period.

Dormice bite into hazel nuts in a characteristic way. The occurrence of dormice nibbled nuts indicates the presence of dormice but cannot be used to determine population size

6.6 **Otter.** Methods for conducting otter surveys are discussed by Jefferies (1986), Kruuk and Conroy (1987) and Kruuk *et al.* (1986). These methods primarily rely on recording otter spraints.

6.7 **Pine marten.** A method for surveying for the presence of pine marten is given in Velander (1983) and Strachen *et al.* (in prep).

6.8 **Red squirrel.** Methods for surveying red squirrels are described by Gurnell (in prep) and Gurnell and Pepper (in prep). Drey counts are best carried out in late winter whilst live trapping can be conducted at most times of the year but tends to be more efficient from February to August/September when food resources are low.

6.9   **Wildcat.** A method for acquiring data on wildcat is described by Easterbee *et al.* (1991).

6.10   **Marine mammals.** Survey methods for seals, dolphins, porpoises and other whales are described in Section 10.

6.11   The above species are those most likely to require detailed studies. If further information on other species is required appropriate survey methods are listed in Table 3. Some of the described methods are appropriate for several species such as trap-mark-recapture techniques and transect surveys. General reviews of these and other survey methods are given in Buckland *et al.* (1993) and Tapper (1992).

Evaluation

6.12   Baseline evaluation of mammal populations presents a variety of problems. For example, some species can be regarded as both vermin or as having nature conservation interest depending on a particular situation. There are also considerable difficulties in applying site specific measures. This is due to the high mobility of many species and the dependency on the wider countryside for their survival. However, a number of species do congregate in large numbers at certain times and the protection of a specific site would therefore be appropriate.

6.13   In evaluating a site, the presence of viable (seasonal or permanent) populations of species included in Schedule 5 of the Wildlife and Countryside Act (Wildlife Order NI) and/or the EC Habitats Directive would suggest a site that is of national (or international) importance. Guidelines for evaluating certain species are given in the handbook for the selection of biological SSSIs (NCC, 1989). In addition, there are other species that are not included in Schedule 5, nor the Directive, which if present on a site might suggest the site is of regional importance. Species of this type could include the polecat, yellow-necked wood mouse and island-races and sub-species of mainly small mammals (NCC, 1989). Reference should also be made to the Red Data Book on mammals (Morris, 1993) and Irish vertebrates (Whilde, 1993) for other species that might fall into this category. Useful information concerning the distribution of all British mammals is given by Arnold (1993).

TABLE 3 ADDITIONAL SURVEY METHODS FOR MAMMALS

Species	Survey method
deer	Buckland (1992), Mitchell and McCowan (1980), Putman (1984), Staines and Ratcliffe (1987)
fat dormouse	Hoodless and Morris (1993)
fox	No particular method devised
grey squirrel	Gurnell (in prep) and Gurnell and Pepper (in prep)
hare	Hutchings and Harris (in prep), Langbein *et al.* (in prep)
hedgehog	No particular method devised
mole	Mead-Briggs and Woods (1973)
mink	Strachan *et al.* (1990)
mountain hare	Angerbjorn (1983), Watson and Hewson (1973)
polecat	No specific method but see Birks (1993)
rabbit	Trout *et al.* (1986), Taylor (1956)
shrew	Gurnell and Flowerdew (1982)
small rodents	Gurnell and Flowerdew (1982), Southern (1973), Twigg (1975a; 1975b)
stoat/weasel	King (1974)
water vole	Lawton and Woodroffe (1991), Stoddart (1970), Woodroffe *et al.* (1990a; 1990b), Strachan and Jefferies (1993)

- Further information on mammals should be presented where protected species may be affected, or where mammals are likely to interact with the operation of a project, or are important for influencing surrounding ecosystems, or where species not covered by existing legislation but perceived as being of local importance, may be affected.

- The species of mammal most likely to require further study are badger, bats, dormouse, otter, pine marten, red squirrel and wildcat. For all species appropriate survey methods, carried out at the correct time of year, should be used.

- Baseline evaluation is complicated by the high mobility of many species and their dependence on the wider countryside rather than a single site. For certain species the SSSI Guidelines (NCC, 1989) allow site evaluation, but some non-protected species may also have regional or local significance.

Amphibians and Reptiles

Survey Criteria

7.1 More detailed information on amphibians and reptiles should be presented under the following circumstances:

When the desk study indicates that the site contains (or has contained in the past) protected species, good assemblages of species, or species on the edge of their range.

For amphibians this means any of the following:

- sites with great crested newt or natterjack toad and in Northern Ireland, smooth newt.
- sites with four or more amphibian species.
- sites in an area where an amphibian is rare or at the edge of its geographical range.

For reptiles this means any of the following:

- sites with sand lizard or smooth snake and in Northern Ireland, common lizard.
- sites with at least three reptile species.
- sites in an area where a reptile is rare or at the edge of its geographical range.

When the extended Phase 1 survey identifies suitable habitat for protected amphibian and reptile species, lying within the known range of the species, but for which no records exist.

Survey Methods for Amphibians

7.2 Survey methods for amphibians are typically based on counts of adults and spawn and are detailed in British Herpetological Society (1990a). The JNCC is in the process of evaluating existing survey methods and will be developing proposals for standardising the way in which they are used. The description of the methods given below is intended to provide a broad indication of the different types of survey methods currently available. Survey methods for amphibians should include searching of suitable terrestrial habitats mapped in the extended Phase 1 survey.

7.3   **Natterjack toad.** Recommended survey techniques are night counts of individuals and spawn counts at breeding pools. These can be carried out from the end of March to mid-July. Several visits will be required during this time. Ideally, data from 5 years are needed to assess the size of a population. However, if age/size data are available, these may provide useful historical information.

7.4   **Newts.** Recommended survey methods are netting, torch counting and bottle trapping; the method used depending on the characteristics of the water body. Counts of newts should be undertaken at night between March and the end of July. To determine the presence of newts at ponds, the 'egg search method' developed by the Lancashire Newt Group (Grayson *et al.*, 1991) can be used. Newts may also be netted during the day when adults are in the ponds. This should be undertaken between February and the end of June for smooth/palmate newts and between February and September for great crested newts. In these surveys the search time should be 15 minutes for sites with less than 50m of water's edge, 30 minutes for those with 50-100m, and pro rata for larger sites (NCC, 1989).

7.5   **Common toad.** Counts in March/April should be carried out during the night for adults and during the day for spawn strings.

7.6   **Common frog.** Spawn clumps should be counted during the day between February and early April.

Survey Methods for Reptiles

Many reptiles and amphibians may use refugia placed on-site in spring or summer, and these can be used to establish species presence (eg slow worms)

Qualitative survey 49
method based on
sightings is available
for Sand lizards

7.7 At present, no standard quantitative technique exists for surveying reptiles, although a qualitative method based on sightings is available (British Herpetological Society, 1990b) and English Nature is carrying out research on developing new survey methods. To establish species presence on a site, refugia (pieces of wood, metal, plastic or fabric beneath which snakes and slow worms hide) can be used, suitable terrestrial habitat should also be mapped. The best time of year for survey is spring and early summer, but some data may be obtained from surveys conducted in the summer and early autumn.

Evaluation

7.8 Survey data should be presented as adults, eggs/spawn or young recorded during the specified time period. Additional information to aid in the evaluation of survey results should include (NCC, 1989):

- assemblage scores;
- population size in local, regional and national terms;
- site location (this referring to isolation and position in a species range as well as the potential function as a corridor between surrounding sites);
- historical perspective in terms of the number of former records.

7.9 A new recording and monitoring scheme for amphibians and reptiles has been established by the Biological Records Centre (ITE) for the JNCC and a distribution atlas for amphibians and reptiles in Great Britain is in press.

Summary

- Further surveys of amphibians and reptiles should be undertaken where sites known to contain protected species, good assemblages of species, or species at the edge of their geographical range, or sites which lie within the known geographical range and which contain suitable amphibian/reptile habitat lie within the impact area.

- Appropriate survey methods should be used for each species.

- Sites should be evaluated in terms of population size and assemblage of species, plus factors such as location in a species range and proximity to other sites.

Introduction

8.1   In addition to the possibility of causing significant impacts on the status or distribution of particular fish species or communities, proposed developments may also affect the exploitation of this group by damaging fisheries.

8.2   In this section only freshwater fish and fisheries are described; marine fish and fisheries are discussed below in Section 10.

Survey Criteria

8.3   More detailed information on fish populations or fisheries should be presented under the following circumstances:

*   when the desk study indicates that the site contains (or has contained in the past) protected species, species known to be in decline, unusual races of species or important fish communities (Maitland,1985).

This means any of the following:

*   Species listed in (a) Schedule 5 of the Wildlife and Countryside Act 1981, (b) Appendix 2 of the Bern Convention, or (c) Annexes II or V of the EC Habitats and species Directive (Boon *et al.,* 1992).

*   Species known to be in decline in the UK at present: allis shad, twaite shad, arctic charr, powan, vendace, pollan, smelt, burbot (Maitland and Lyle, 1991).

*   Unusual races, for example, 'landlocked' river lamprey, in Loch Lomond, spineless three-spined stickleback in the Hebrides (Maitland and Lyle, 1991).

*   Important fish communities, i.e. those with unusual assemblages of species (eg Llyn Tegid, Haweswater, Loch Lomond, Loch Eck) or with communities which are believed to be pristine (e.g. Langavat) (Maitland and Lyle, 1991).

- When the desk study indicates that a fishery is likely to be affected, for example by the interruption of fish migration, and damage to spawning grounds.

Survey Methods

8.4   The need for survey and the methods required will vary greatly depending on the amount of information already available from various sources, the species (and life history stage) of fish involved and the type of habitat to be sampled.

8.5   Some information may be available from surveys previously carried out by other workers/organisations, for example the National Rivers Authority (England and Wales), the River Purification Boards and District Salmon Fishery Boards (Scotland), the Fishery Boards (Northern Ireland), government and independent scientists.

8.6   Where a survey is necessary, appropriate methods vary greatly according to fish species, life stage and aquatic habitat. The description of methods given below is intended to provide a broad indication of the different types of survey methods available. Fuller details are given in various publications, for example Bagenal (1978), Nielsen *et al.* (1983) and Maitland (1990).

Electro-fishing can be used for monitoring of fish populations

- Streams. In most cases, quantitative electro-fishing is the most accepted method, though drift nets for larvae and traps and counters for migratory species may also provide reliable data.

- Rivers. Electro-fishing is more difficult in larger rivers and here seine netting, traps and tow nets (for larvae) are more effective. However, in large running waters, quantitative estimates are difficult to obtain.

- Ponds. Population estimates are best obtained from mark-recapture methods, using fish caught by seine netting or trapping.

- Lakes. Population estimates can be obtained from mark-recapture methods, using fish caught by seine netting or trapping. However, in larger lakes, total estimates are very difficult without enormous effort and here only relative numbers may be feasible using standard mixed-mesh gill nets, traps and other methods. Ichthyoplankton nets and other specialised samplers may be required to sample pelagic larvae.

- For some species, estimates of the numbers of eggs (eg smelt) or nests (eg salmonid redds) may be obtained by direct counts immediately after spawning.

8.7   It is important, in addition to obtaining data on fish numbers, to assess the loss of different habitats which may be important to the various life stages of fish. This information can then be used to develop a 'no net loss of fish habitat' strategy (Department of Fisheries and Oceans, 1986).

Evaluation

8.8   Quantitative survey data should be presented as eggs/nests, young (larvae and juveniles) or adults estimated per unit area at a particular time. With fish counters or traps for migratory species, the data can be expressed as numbers passing per unit of time.

8.9   Relative abundance is usually expressed as Catch Per Unit Effort (CPUE) and it is important to state clearly which method was used (eg standard 50m Lundgrens mixed-mesh multifilament gill net per 24 hours, etc).

8.10   In addition, to aid the evaluation of survey results, the following are also desirable:

- Population size in local, regional and national terms.
- Site location and size - this is especially important for

running waters where normally the data have been obtained from one particular stretch.

- Historical perspective in terms of any previous data available.

- Where fisheries are involved, previous catch data for the water, ideally in terms of CPUE.

- Where data are likely to be required in the future (eg to assess the actual impact of a development) a suitable system of monitoring fish numbers should be established using standard methods (Maitland and Lyle 1992).

Summary

- Surveys should be undertaken where the site contains (or has contained in the past) protected species, species known to be in decline, unusual races of species or important fish communities, or valuable fisheries.

- Appropriate methods should be used, but these vary greatly according to fish species, life stage and aquatic habitat.

- Data should be evaluated on the basis of fully comparable quantitative or relative information.

Terrestrial and Aquatic Invertebrates

Introduction

9.1   The British invertebrate fauna is extremely large; the insect fauna alone comprises some 22 500 species whose biology and habitat requirements are, for the most part, either unknown or incompletely understood. Notwithstanding this, there is a vast array of literature on both terrestrial and aquatic invertebrates. Some of the main sources of information are outlined below.

9.2   Ball (1986) has identified uncommon invertebrate species which are classed as Nationally Notable (= Nationally Scarce). This category is subdivided into Notable A (Na) and Notable B (Nb) based upon a species known distribution, although in some recent reviews these have been combined into a single class of Notable (N) species. More comprehensive reviews of scarce and threatened species have been published recently for the following orders of insect: Coleoptera (Hyman and Parsons, 1992; 1994); Diptera Pt.1 (Falk,1991a);   Hemiptera   (Kirby,   1992a);   Hymenoptera (Falk,1991b); Lepidoptera: Pyralidae (Parsons, 1993); Trichoptera (Wallace, 1991); Ephemeroptera and Plecoptera (Bratton, 1990); and Neuroptera (Kirby, 1991). A comprehensive review of scarce spiders is given in Merrett (1990).

9.3   Red Data Book (RDB) insects in Great Britain are described in Shirt (1987). This provides species accounts for all Endangered (RDB 1) and Vulnerable (RDB 2) species, but only lists those classed as Rare (RDB 3). Details of the RDB status of invertebrates, other than insects, are provided by Bratton (1991).

9.4   Britain has an international obligation to protect certain species of invertebrate under the terms of both the EC Habitats Directive, and the Bern Convention (see Appendix 3). These include species such as the stag beetle which, in Britain, is only regarded as Nationally Scarce. Species of invertebrate are also protected under the Wildlife and Countryside Act, which covers selected RDB species including 11 Lepidoptera (4 butterflies and 7 moths), 7 Coleoptera, 1 Homoptera, 3 Orthoptera, 1 Odonata, 2 Araneae, and 3 non-marine Mollusca (see Appendix 3) in addition to species protected against trade.

9.5   More detailed information on terrestrial and aquatic invertebrates should be presented under the following circumstances:

•   For all sites where the development will have a direct or indirect impact on the water quality of rivers or still waters, a baseline survey of the aquatic invertebrate fauna should be undertaken unless adequate data already exist.

•   When this initial water quality invertebrate survey sample achieves any of the following values equivalent to the top 10% of samples from the 1990 survey of the UK

-        26 or more families of invertebrates
-        BMWP score of 150 or greater
-        ASPT score of 6.48 or greater

then the samples taken should be analysed to species level wherever practical.

•   When the desk study reveals that a RDB invertebrate has previously been recorded from the site, and the extended Phase 1 survey indicates that suitable habitat for it still exists within the impact area. If the presence of the species is not detected on the site, the likelihood of its continued presence should be assessed in terms of the survey coverage and species ecology.

•   When the desk study indicates that Nationally Scarce species of invertebrate are present on the site, or RDB or Nationally Scarce species of invertebrate occur near the site in habitats similar to those present within the study area. In the latter case the relevant on-site habitats may need to be surveyed to assess their value for invertebrates in comparison with the nearby areas of known invertebrate value.

•   When the extended Phase 1 survey identifies features or habitats of significant value to invertebrates (Kirby, 1992b), for example, dying timber, ancient woodland and fens.

•   When the desk study reveals that the site qualifies as a dragonfly key site (SSSI Citation), as this is a good indicator for quality habitat for invertebrates.

9.6   Appropriate survey methods for invertebrates depend upon the target groups and habitats under consideration. Most methods likely to be employed are described in Southwood (1977) whilst reference should also be made to Brooks (1993a) which contains advice on choice of target groups, survey methods, points to cover in the survey report and legal considerations.

9.7   Surveys for terrestrial insects and most other invertebrates should be carried out between May and September and optimally three periods of sampling should be carried out in early, mid- and late season.

9.8   Because of the vast number of species and the range of different invertebrate organisms involved, field surveys should initially be restricted to a few target groups which are characteristic of the habitats present on-site and for which good biological/ecological data and identification keys are readily available. Suggested groups include: Carabidae (ground beetles); Lepidoptera (butterflies and moths); Orthoptera (crickets and grasshoppers); Syrphidae (hoverflies); and adult Odonata (dragonflies and damselflies). The target groups used will, however, vary according to the habitat type being investigated.

9.9   Any RDB species recorded during surveys should be confirmed by a recognised expert in the group, or a national reference collection. The status of RDB and Nationally Scarce species is being constantly updated as more detailed information becomes available concerning their biology, habitats and distribution. Care must therefore be taken to refer to the most recent review available.

Survey Methods for Aquatic Invertebrates

Conservation Surveys

9.10   There are various methods for assessing the conservation value of water bodies based on aquatic invertebrate sampling. For example, ditch systems can be evaluated using methods outlined by Drake (1987; 1988; 1991). Pond Action (1993) describes a method for assessing ponds based on the identification of macro-invertebrates and plants. The conservation value of ponds can also be evaluated on the basis of their water beetle communities (Foster and Eyre, 1992). Dragonfly species assemblages also provide a good indication

58   Kick sampling for
invertebrates to
assess water quality

of the conservation value of a water body. Methods for monitoring
adult dragonflies are described by Moore and Corbet (1990) and
Brooks (1993b).

Water Quality Surveys

9.11   Several biotic indices of varying complexity have been developed to assess water quality (Hellawell, 1978; 1986) but the most
widely used at present are the BMWP score (Anon, 1981) and the
Environmental Quality Index from the Rivers Invertebrate
Prediction and Classification Scheme (RIVPACS) (Wright *et al.*,
1989). These indices have been designed to monitor the water
quality of freshwater, but they also reflect general environmental and
habitat quality.

9.12   Sampling methods should be standardised to allow comparisons between sampling sites and over time. For example, during surveys of shallow rivers a standard number of kick samples should be taken across the riverbed at each sampling station, and all habitat types (eg riffles, amongst vegetation) should be sampled at each. For deeper, slow-flowing rivers or standing waters a standard number of net sweeps, preferably through aquatic vegetation to maximise the numbers of invertebrates sampled, should be made at each sampling station or the period of sampling restricted by set time limits. Sampling stations should be selected at regular intervals along the bank and a standard 250 mm square frame net with 1mm mesh net bag should be used for the sampling. For deeper water other methods such as colonisation samplers, corers, grabs or air-lift samplers will be necessary (Hellawell, 1978; Lincoln and Sheals, 1979).

9.13   All invertebrates should be identified to species level where practicable. Species level identification is recommended where identification keys are available, particularly for sensitive indicator groups: Ephemeroptera (Elliott *et al.*, 1988); Plecoptera (Hynes, 1977); Trichoptera (Edington and Hildrew, 1981; Wallace *et al.*, 1990); Odonata (Hammond, 1983; Miller, 1987; Askew, 1988); Coleoptera (Friday, 1988). This will also allow the identification of RDB species (Shirt, 1987) or locally uncommon species (Bratton, 1990; Wallace, 1991; Hyman and Parsons, 1992). A useful overview of methods and their application is provided by Team (in press).

9.14   Aquatic invertebrates can be sampled throughout the year and preferably on a seasonal basis. Adult dragonflies are best sampled between late May and the end of August, preferably at monthly intervals.

Evaluation

9.15   All notable and RDB species should be detailed with an assessment of their abundance on the site. If possible their status should be compared with existing records for the study area, district or county/region.

Summary

*   Surveys of freshwater aquatic invertebrates should be under-
        taken whenever a proposed development is predicted to have
        an impact on freshwater quality. When this initial survey
        reveals 26 or more families present, or a BMWP score >150

or an ASPT score >6.48 then the sample should be analysed to species level where practicable. Further surveys of terrestrial invertebrates should be under-taken where RDB or Nationally Scarce species may be affected or where habitats similar to nearby areas of known invertebrate interest lie within the impact area.

• All surveys should be undertaken at the correct time of year using techniques appropriate to the species and/or habitat type under consideration.

• Target groups can be used as bio-indicators to characterise the main invertebrate communities under investigation.

• The identity of any RDB species must be confirmed.

• The status of all notable and RDB species should be compared, where possible, to existing data on the study area, district or county/region.

PART FOUR

Marine and estuarine habitats
and organisms

Marine and Estuarine Habitats and Organisms

Introduction

10.1   This section covers the requirements for more detailed surveys in marine and estuarine systems. Since it covers all ecological groups, criteria to trigger more detailed surveys are not given.

10.2   Areas of impact in marine and estuarine areas are potentially large. Also, repeated sampling is often required because of variation between seasons and tidal cycles. For these reasons, as well as to avoid unnecessary cost and effort, existing data obtained by consultation should be used where possible.

Survey criteria

10.3   It is possible to give general guidance on situations where more detailed ecological information should be presented in the ES. These are as follows:

•   When the desk study reveals that any part of a SSSI, NNR, Marine Nature Reserve, proposed or designated SPA, SAC or RAMSAR site, or an area listed as important for marine wildlife (English Nature, 1993), lies within the impact area.

•   When the desk study or Phase 1 survey indicates that any protected species, any areas of designated or known importance as a nursery area for fish, or any significant commercial or recreational fishery may be affected.

Survey Methods

10.4   **Surveys.** Should more detailed information be needed which cannot be obtained by consultation, suitable survey methods are described below. Only the groups appropriate to the interest of the site or type of impact would need to be surveyed. Environmental Advisory Unit (1993) gives a more detailed review of standard ecological survey methods in these habitats.

10.5   **Vegetation.** For some marine and estuarine habitats, there are particular methods which can be used in addition to, or instead of, the National Vegetation Classification. They can also provide data in a form comparable to the relevant national database. These include the national shingle survey method (Sneddon and Randall, 1993), which used a different and more detailed classification to that of the NVC and the salt marsh survey of Great Britain (Burd, 1989) which used mapping of communities by dominant species.

10.6   Recommended survey techniques for subtidal vegetation are essentially the same as for invertebrates (see below).

10.7   **Birds.** see Section 5.

10.8   **Mammals.** A number of techniques for surveying cetaceans (dolphins, porpoises and other whales) have been devised and are reviewed in Hiby and Hammond (1989) and Hammond (1987). For the study of coastal dolphins a useful method is to use land-based observers as described by Hammond and Thompson (1991).

Seal hauling ground on the Lincolnshire coast (Donna Nook)

10.9   Surveys of seals are principally conducted during the breeding and moulting seasons when the animals are land-based. Survey techniqus include aerial and time-lapse photography and walked or boat based transects. Discussions of these methods are given in Hiby *et al.* (1988), Thompson and Harwood (1990) and Ward *et al.* (1987).

10.10 **Invertebrates.** Habitats to be sampled for marine/estuarine invertebrates are intertidal and subtidal substrata for benthic species and coastal waters for pelagic species. Sampling techniques typically include cores, quadrats, grabs, dredges, nets, trawls and remote-controlled cameras. Often a combination of sampling methods will be needed to sample the range of habitats on a section of shoreline. Methods will often require sampling along transects or within zones, to reflect differences in elevation and frequency/depth of submergence by tidal waters. Possible survey and sampling methods for benthic and pelagic invertebrates are summarised in Baker and Wolff (1987), Holme and McIntyre (1984) and MAFF (1993).

10.11 Standard Phase 2 survey methods for marine/estuarine invertebrates and their habitats have been developed by the Marine Nature Conservation Review (MNCR). Methods involve completing a littoral and/or sublittoral record for the whole site, in addition to records for each habitat, including species abundance scores. These methods are summarised in Hiscock (1990), but this is now somewhat out-of-date and the MNCR should be contacted directly for guidance.

10.12 **Fish.** Survey methods will depend on the site and species under study. In most circumstances, sampling should be undertaken on a seasonal (quarterly) basis. Reference to local commercial fisheries (ie landings) will indicate the most abundant commercial species and the fishing gear which it is most appropriate to use. For juvenile and small species, beach seine nets may be used in shallow water with a clear seabed, or in deeper water a small beam trawl, such as that used routinely for young flatfish surveys (or commercial shrimp fishing), may be appropriate. Further advice may be obtained from the Directorate of Fisheries Research (DFR) in England and Wales or the SOAFD Marine Laboratory in Aberdeen.

Evaluation

10.13 For each of the described habitat types and taxa, the evaluation of collated baseline data should follow the recommendations given in the relevant Sections 4-9.

- Further surveys of marine and estuarine habitats should be undertaken where the impact area includes a statutory site of nature conservation interest, or is considered to be important for marine wildlife, or where protected species, or an important nursery area for fish, or a fishery may be affected.

- Appropriate survey methods should be used for each ecological group.

- Each habitat type and taxa should be evaluated according to the information proposed in the relevant section of the guidelines.

Abbreviations

ASPT	Average Score per Taxon
BMWP	Biological Monitoring Working Party
BS7750	British Standard: specification for Environmental Management Systems (1994)
CPUE	Catch Per Unit Effort
CIMAH	Control of Industrial Major Accident Hazard Registration (1984)
DFR	Directory of Fisheries Research
EAs	Environmental Assessments
EC	European Council
EIA	Environmental Impact Assessment
IPC	Integrated Pollution Control
ISR	Invertebrate Site Register
ITE	Institute of Terrestrial Ecology
IUCN	International Union for Conservation of Nature
JNCC	Joint Nature Conservation Committee
MNCR	Marine Nature Conservation Review
NCC	Nature Conservancy Council
NCV	National Vegetation Classification
NNR	National Nature Reserve
RDB	Red Data Book
RIVPACS	Rivers Invertebrate Prediction and Classification Scheme
SAC	Special Area of Conservation
SINC	Site of Importance for Nature Conservation
SOAFD	Scottish Office Agriculture & Fisheries Department
SPA	Special Protection Area
SSSI	Site of Special Scientific Interest

References

Alcock, M.R. and Palmer, M. A. (1985) A Standard Method for the Survey of Ditch Vegetation. CST Report No. 37, Nature Conservancy Council, Unpublished.

Angerbjorn, A. (1983) The Reliability of Pellet Counts as Density Estimates of Mountain Hares. *Finnish Game Research*, 41,433-488.

Anonymous (1988) Fish Populations Shallow Rivers and Streams. Standing Council Analyst, Department of the Environment, HMSO, London.

Anonymous (1981) River Quality: the 1980 Survey and Future Outlook. National Water Council.

Arnold, H.R. (1993) Atlas of Mammals in Britain. Institute of Terrestrial Ecology Research Publication, No.6, HMSO, London.

Askew, R.R. (1988) The Dragonflies of Europe. Harley Books, Colchester.

Bagenal, T.B. (1978) Methods for Assessment of Fish Production in Freshwaters. Blackwell Scientific Publications, Oxford.

Baker, J.M. and Wolff, W.J. (Eds) (1987) Biological Surveys of Estuaries and Coastal Habitats. Estuarine and Brackish Water Sciences Association Handbook, Cambridge University Press, Cambridge.

Ball, S.G. (1986) Terrestrial and Freshwater Invertebrates with Red Data Book, Notable or Habitat Indicator Status. Invertebrate Site Register Report (CTD Report No.637), No.66, Nature Conservancy Council, Peterborough.

Batten, L.A., Bibby, C.J., Clement, P., Elliott, G.D. and Parker, R.F. (1990) Red Data Birds in Britain. Poyser, London.

Bibby, C.J., Burgess, N.D. and Hill, D.A. (1992) Birds Census Techniques. Poyser, London.

Birks, J. (1993) The Return of the Polecat. *British Wildlife*, 5, 16-25.

70   Boon, P.J., Morgan, D.H.W. and Palmer, M.A. (1992) Statutory Protection of Freshwater Flora and Fauna in Britain. *Freshwater Forum*, 2, 91-101.

Box, J.D. and Forbes, J.E. (1992) Ecological considerations in the Environmental Assessment of Road Proposals. *Highways and Transportation* <u>39</u>, 16-22.

Bratton, J.H. (1990) A Review of the Scarcer Ephemeroptera and Plecoptera of Great Britain. Research and Survey in Nature Conservation, No.29, Nature Conservancy Council, Peterborough.

Bratton, J.H. (Ed) (1991) British Red Data Books: 3. Invertebrates other than Insects. Joint Nature Conservation Committee, Peterborough.

Bright, P.W., Mitchell, P. and Morris, P.A. (in press) Dormouse Distribution, Survey Techniques, Insular Ecology and Selection of Sites for Conservation. *Journal of Applied Ecology.*

Bright P.W. and Morris P.A. (1990) A Practical Guide to Dormouse Conservation. Occasional Publication, No. 11, Mammal Society London.

British Herpetological Society (1990a) Surveying for Amphibians.

British Herpetological Society (1990b) Save our Reptiles.

British Trust for Ornithology (1982) Waterways Bird Survey: Instructions. British Trust for Ornithology, Thetford.

British Trust for Ornithology (1983) Common Bird Census: Instructions. British Trust for Ornithology, Thetford.

British Trust for Ornithology (1992) National Low Tide Counts: Detailed Instructions. British Trust for Ornithology, Thetford.

British Trust for Ornithology (undated (a)). Instructions to Counters: Breeding Waders of Wet Meadows Survey. British Trust for Ornithology, Thetford.

British Trust for Ornithology (undated (b)). Wetland Bird Survey: Instructions. British Trust for Ornithology, Thetford.

Brooks, S.J. (1993a) Joint Committee for the Conservation of British Invertebrates: Guidelines for invertebrate site surveys. *British Wildlife,* 4 (5): 283-286.

Brooks, S.J. (1993b) Review of a Method to Monitor Adult Dragonfly Populations. *Journal of the British Dragonfly Society,* 9 (1), 1-4.

Brown, A.F. and Shepherd, K.B. (1993) A Method for Censusing Upland Breeding Waders. *Bird Study,* 40, 189-195.

Buckland, S.T., Anderson, D.R., Burnham, K.P. and Laake, J.L. (1993) Distance Sampling, Estimating Abundance of Biological Populations. Chapman and Hall, London.

Buckland, S.T. (1992) Review of Deer Count Methodology. Scottish Office, Edinburgh.

Burd, F. (1989) The Saltmarsh Survey of Great Britain: An Inventory of British Saltmarshes. Research and Survey in Nature Conservation, No. 17, Nature Conservancy Council, Peterborough.

Clements, D.K. and Tofts, R.J. (1992) Hedgerow Evaluation and Grading System (HEGS) - A Methodology for the Ecological Surveys, Evaluation and Grading of Hedgerows (Test Draft). Countryside Planning and Management, Cirencester.

Coles, T.F. and Smith, A. (1993) From natural history ramblings to scientific assessments - improving standards for ecological inputs to environmental assessments. In Proceeding of the 5th Annual Advances in Environmental Impact Assessment Conference, IBC Technical Services, London.

Collis, I. and Tyldesley, D. (1993) Natural Assets: Non-Statutory Sites of Importance for Nature Conservation. Local Government Nature Conservation Initiative, Hampshire County Council.

Commission of the European Communities (1979) Council Directive 79/409/EEC on the Conservation of Wild Birds. *Official Journal of the European Communities,* L103.

Commission of the European Communities (1985) Council Directive 85/337/EEC on the Assessment of the Effects of Certain Public and Private Projects on the Environment. *Official Journal of the European Communities,* L175, 40-48.

Commission of the European Communities (1991) CORINE Biotopes: The Design Compilation and Use of an Inventory of Sites of Major Importance for Nature Conservation in the European Community, Commission of the European Communities, Luxembourg.

Commission of the European Communities (1992) Council Directive 92/43/EEC on the Conservation of Natural Habitats and of Wild Flora and Fauna. *Official Journal of the European Communities*, L206, 7-49.

Corbet, G.B. and Harris, S. (1991) The Handbook of British Mammals, (third edition). Blackwell Scientific Publications, London.

Countryside Commission (1990) Countryside and Nature Conservation Issues in District Local Plans. Countryside Commission, Cheltenham.

Countryside Commission, English Heritage and English Nature (1993) Conservation Issues in Strategic Planning. Countryside Commission, Northampton.

Cowx, I.G. (Ed) (1990a) Developments of Electric Fishing. Fishing News Books.

Cowx, I.G. (Ed) (1990b) Catch Effort Sampling Strategy. Fishing News Books.

Cresswell, P., Harris, S. and Jefferies, D.J. (1990) The History, Distribution, Status and Habitat Requirements of the Badger in Britain. Nature Conservancy Council, Peterborough.

Curtis, T.G.F. and McGough, H.N. (1988) Irish Red Data Book:1. Vascular Plants. Stationery Office, Dublin.

Department of the Environment (Welsh Office) (1987) Joint Circular 27/87 (52/87). Nature Conservation. HMSO, London.

Department of the Environment (1988) Planning Policy Guidance Note 1. General Policy and Principles. HMSO, London.

Department of the Environment (1991) Circular 16/91. Planning Obligations. HMSO, London.

Department of the Environment (Welsh Office) (1992a) Joint Circular 1/92. Planning Controls over Sites of Special Scientific Interest. HMSO, London.

Department of the Environment (1992b) Planning Policy Guidance Note 12. Development Plans and Regional Planning Guidance. HMSO, London.

Department of the Environment (1994) Planning Policy Guidance Note 9: Nature Conservation. HMSO, London.

Department of the Environment (1993) Environmental Appraisal of Development Plans: A Good Practice Guide. HMSO, London.

Department of Fisheries and Oceans (1986) Policy for the Management of Fish Habitat. Department of Fisheries and Oceans, Ottawa.

Department of Trade and Industry (1992) Guidelines for the Environmental Assessment of Cross-Country Pipelines. HMSO, London.

Department of Transport (1993) Design Manual for Roads and Bridges Vol. 11. Environmental Assessment. Department of Transport, London.

Donn, S. and Wade, M. (1994) UK Directory of Ecological Information. Packard Publishing Ltd, Chichester.

Drake, C.M. (1987) Dragonflies on the Gwent and Somerset Levels and Moors. *Journal of the British Dragonfly Society*, 3(1), 1-4.

Drake, C.M. (1988) Diptera from the Gwent Levels, South Wales. *Entomologist's Monthly Magazine*, 124, 37-44.

Drake, C.M. (1991) Ephemeroptera and Plecoptera in Freshwater and Brackish Ditch Systems on British Grazing Marshes. *Entomologist's Gazette*, 42, 45-59.

Earll, R. (1992) The SEASEARCH Habitat Guide - An Identification Guide to the Main Habitats Found in the Shallow Seas Around the British Isles. Marine Conservation Society and Joint Nature Conservation Committee.

Easterbee, N., Hepburn, L.V. and Jefferies, D.J. (1991) Survey of the Status and Distribution of the Wildcat in Scotland 1983-1987. Nature Conservation Council for Scotland, Edinburgh.

Edington, J.M. and Hildrew, A.G. (1981) Caseless Caddis Larvae of the British Isles. Freshwater Biological Association, Ambleside.

Elliott, J.M., Humpesch, U.H. and Macan, T.T. (1988) Larvae of the British Ephemeroptera. Freshwater Biological Assocation, Ambleside.

English Nature (1993) Managing England's Marine Wildlife. Draft for Consultation.

English Nature (1992) Strategic Planning and Sustainable Development (Consultation Paper). English Nature, Peterborough.

English Nature (1994) Nature Conservation in Environmental Assessment. English Nature, Peterborough.

Environmental Advisory Unit (1993) Review of Nature Conservation Survey Methodologies. R & D Note 107. National Rivers Authority.

Falk, S. (1991a) A Review of the Scarce and Threatened Flies of Great Britain (Part 1). Research and Survey in Nature Conservation, No.39, Nature Conservancy Council, Peterborough.

Falk, S. (1991b) A Review of the Scarce and Threatened Bees, Wasps and Ants of Great Britain. Research and Survey in Nature Conservation, No.35, Nature Conservancy Council, Peterborough.

Forbes, J. and Heath, D. (1990) The Ecological Impact of Road Schemes. Nature Conservancy Council and Department of Transport.

Foster, G.N. and Eyre, M.D. (1992) Classification and Ranking of Water Beetle Communities. UK Nature Conservation, No.1, Nature Conservancy Council, Peterborough.

Friday, L.E. (1988) A Key to the Adults of British Water Beetles. Field Studies, 7(1) 1-152.

Gibbons, D.W., Reid, J.B. and Chapman, R. (Eds) (1993) The New Atlas of Breeding Birds in Britain and Ireland. T and A D Poyser.

Gibbons, J.E. (1975) The Flora of Lincolnshire. Lincolnshire Natural History Brochure, No.6, Lincolnshire Naturalist's Union, Lincoln.

Grayson, R.F., Parker, R. and Mullaney, A.S. (1991) Atlas of the Amphibians of Greater Manchester County and New Criteria for Appraising UK Amphibian Sites. *Lancashire Wildlife Journal*, No. 1.

Gurnell, J. (in prep.) The Mammal Society, London.

Gurnell J. and Pepper H. (in prep.) Forestry Commission, Edinburgh.

Gurnell, J. and Flowerdew, J.R. (1982) Live Trapping Small Mammals. A Practical Guide. The Mammal Society, Reading.

Hammond, C.O. (1983) The Dragonflies of Great Britain and Ireland. Harley Books, Great Horkesley.

Hammond, P.S. (1987) Techniques for Estimating the Size of Whale Populations. Symposium of the Zoological Society, London, 58, 225-245.

Hammond, P.S. and Thompson P.M. (1991) Minimum Estimate of the Number of Bottlenose Dolphins (*Tursiops truncatus*) in the Moray Firth, N.E.Scotland. *Biological Conservation*, 56, 79-87.

Harding, P.T. (1993) Current Atlases of Flora and Fauna in the British Isles. Biological Records Centre, Huntingdon.

Harris, S., Cresswell, P. and Jefferies, D.J. (1989) Surveying Badgers. The Mammal Society, London.

Hawksworth, D.L. and Rose, F. (1976) Lichens as Pollution Monitors. Studies in Biology, 66, Arnold, London.

Hellawell, J.M. (1978) Biological Surveillance of Rivers. Water Research Centre.

Hellawell, J.M. (1986) Biological indicators of freshwater pollution and environmental management. Elsevier Applied Science, London.

Helliwell, D.R. (1973) Priorities and Values in Nature Conservation. *Journal of Environmental Management*, 1, 85-127.

76

Helliwell, D.R. (1973) Priorities and Values in Nature Conservation. *Journal of Environmental Management*, 1, 85-127.

Hiby A.R., Thompson D. and Ward A.J. (1988) Census of Grey Seals by Aerial Photography. *Photogrammetric Record*, 12, 589-594.

Hiby A.R. and Hammond P.S. (1989) Survey Techniques for Estimating Abundance of Cetaceans. Report to the International Whaling Commission (Special Issue II), 47-80.

Hill, M.O. Bourn, R. and Walker, C.J. (1992) Quantative Fish Stock Assessment Choices, Dynamics. Chapman and Hall, London.

Hill, M.O. (1993) TABLEFIT Version 0.0 - For Identification of Vegetation Types. Institute of Terrestrial Ecology, Huntingdon.

Hill, M.O., Preston, C.D. and Smith, A.J.E. (Eds) (1991) Atlas of the Bryophytes of Great Britain and Ireland. Volume 1, Liverworts Hepaticae and Anthocerotae. Harley Books, Colchester.

Hill, M.O., Preston, C.D. and Smith, A.J.E. (Eds) (1992) Atlas of the Bryophytes of Great Britain and Ireland. Volume 2, Mosses (except Diplolepideae). Harley Books, Colchester.

Hill, M.O., Preston, C.D. and Smith, A.J.E. (Eds) (1994) Atlas of the Bryophytes of Great Britain and Ireland. Volume 3, Mosses (Diplolepideae). Harley Books, Colchester.

Hiscock, K. (1990) Marine Nature Conservation Review: Methods. CSD Report No. 1072. MNCR Occasional Report MNCR/OR/05. Nature Conservancy Council, Peterborough.

Hodgetts, N.G. (1992) Guidelines for Selection of Biological SSSIs: Non-Vascular Plants. Joint Nature Conservation Committee, Peterborough.

Holme, N.A. and McIntyre, A.D. (1984) Methods for the Study of Marine Benthos. Blackwell Scientific Publications Oxford.

Holmes, N. (1987) Typing of British Rivers According to their Flora. HMSO. Also published in 1983 in Focus on Nature Conservation, No. 4, Nature Conservation Council, London.

Hoodless, A. and Morris, P.A. (1993) An Estimation of the Population Density of the Fat Dormouse (*Glis glis*). *Journal of Zoology*, 230, 337-340.

Hutchings, M. and Harris, S. (in prep.) Joint Nature Conservation Committee, Peterborough.

Hyman, P.S. and Parsons, M.S. (1992) A Review of the Scarce and Threatened Coleoptera of Great Britain (Part 1). UK Nature Conservation, No.3, Joint Nature Conservation Committee, Peterborough.

Hyman, P.S. and Parsons, M.S. (1994) A Review of the Scarce and Threatened Coleoptera of Great Britain (Part 2). UK Nature Conservation, No.12, Joint Nature Conservation Committee, Peterborough.

Hynes, H.B.N. (1977) A Key to the Adults and Nymphs of British Stoneflies (Plecoptera). Freshwater Biological Associations, Ambleside.

James, P.W., Hawksworth, D.L. and Rose, F. (1977) Lichen Communities in the British Isles: A Preliminary Conspectus. In M.R.D. Seaward (Ed) Lichen Ecology, 295-413, Academic Press, London.

Jefferies, D.J. (1986) The Value of Otter *Lutra lutra* Surveying using Spraints: An Analysis of its Successes and Problems in Britain. *Journal of the Otter Trust*, 1985, 1,25-32.

Joint Nature Conservation Committee (1993) Handbook for Phase 1 Habitat Survey: A Technique for Environmental Audit. Joint Nature Conservation Committee, Peterborough.

King, C.M. (1974) A System of Trapping and Handling Live Weasels in the Field. *Journal of Zoology*, 171, 255-264.

Kirby, K.J. (1988) A Woodland Survey Handbook, Research and Survey in Nature Conservation, No. 11, Nature Conservancy Council, Peterborough.

Kirby, P. (1991) A Review of the Scarcer Neuroptera of Great Britain. Research and Survey in Nature Conservation, No.34, Joint Nature Conservation Committee, Peterborough.

78

Kirby, P. (1992a) A Review of the Scarce and Threatened Hemiptera of Great Britain. UK Nature Conservation, Joint Nature Conservation Committee, Peterborough.

Kirby, P. (1992b) Habitat Management for Invertebrates: A practical handbook. RSPB, Sandy.

Kruuk, H. (1978) Spacial Organisation and Territorial Behaviour of the European Badger *Meles meles. Journal of Zoology*, 184, 1-19.

Kruuk, H. and Conroy, J.W.H. (1987) Surveying Otter *Lutra lutra* Populations: A Discussion of Problems with Spraints. *Biological Conservation*, 41,170-183.

Kruuk H., Conroy, J.W.H., Glimmerveen, V. and Oukerkerk, E.J. (1986) The Use of Spraints to Survey Populations of Otter *Lutra lutra. Biological Conservation*, 35,187-194.

Langbein, J., Hutchings, M., Harris, S., Stoate, C., Tapper, S. and Wray, S.,(in prep.) Techniques for Assessing the Abundance of Brown Hares, *Lepus europaeus.*

Lawton, J.H. and Woodroffe, G. (1991) Habitat and the Distribution of Water Voles. Why are there Gaps in the Species Range? *Journal of Animal Ecology*, 60,79-91.

Lincoln, R.J. and Sheals, J.G. (1979) Invertebrate Animals, Collection and Preservation. British Museum (Natural History), London.

Lloyd, C., Tasker, M.L. and Partridge, K. (1991) The Status of Seabirds in Britain and Ireland. T. and A.D. Poyser.

MAFF Directorate of Fisheries Research (1993) Analysis and interpretation of benthic community data at UK sewage sludge disposal sites. Aquatic Environment Monitoring Report No.37, MAFF Directorate of Fisheries Research, Lowestoft.

Maitland, P.S. (1972) A Key to the Freshwater Fishes of the British Isles with Notes on their Distribution and Ecology. Scientific Publication of the Freshwater Biological Association, 27.

Maitland, P.S. (1985) Criteria for the Selection of Important Sites for Freshwater Fish in the British Isles. *Biological Conservation*, 31, 335-353.

Maitland, P.S. (1990) Biology of Fresh Waters. Blackie, Glasgow.

Maitland, P.S. and Lyle, A.A. (1991) Conservation of Freshwater Fish in the British Isles: The Current Status and Biology of Threatened Species. *Aquatic Conservation*, 1, 25-54.

Maitland, P.S., and Lyle, A.A. (1992) Conservation of Freshwater Fish in the British Isles: Proposals for Management. *Aquatic Conservation*, 2, 165-183.

Malloch, A.J.C. (1991) MATCH (Version 1.3) A computer program to aid the assignment of vegetation data to the communities and subcommunities of the National Vegetation Classification. University of Lancaster.

Margules, C.R. and Usher, M.B. (1981) Criteria Used in Assessing Wildlife Conservation Potential: A Review. *Biological Conservation*, 21, 79-109.

Mead-Briggs, A.R. and Woods, J.A. (1973) An Index of Activity to Assess the Reduction in Mole Numbers Caused by Control Measures. *Journal of Applied Ecology*, 10, 837-845.

Merrett, P. (1990) A Review of the Nationally Notable Species of Spiders of Great Britain. CSD Contract Report, No.127, Nature Conservancy Council, Peterborough.

Miller, P.L. (1987) Dragonflies. Naturalists' Handbooks 7. Cambridge University Press, Cambridge.

Mitchell, B. and McCowan, D. (1980) Estimating and Comparing Population Densities of Red Deer (*Cervus elaphus* L.) in Concealing Habitats. Annual report of the Institute of Terrestrial Ecology 1979, 7-13.

Moore, N.W. and Corbet, P.S. (1990) Guidelines for monitoring dragonfly populations. *Journal of the British Dragonfly Society*, 6 (2) 21-23.

Morris, P. (1993) Red Data Book for British Mammals. The Mammal Society, London.

Nature Conservancy Council (1986) Nature Conservation Guidelines for Onshore Oil and Gas Development. Nature Conservancy Council, Peterborough.

80

Nature Conservancy Council (1987) The Bat Worker's Manual. Nature Conservancy Council, Peterborough.

Nature Conservancy Council (1989) Guidelines for Selection of Biological SSSIs. Nature Conservancy Council, Peterborough.

National Rivers Authority (1992) River Corridor Surveys: Methods and Procedures. Conservation Technical Handbook No.1, National Rivers Authority, Reading.

Nielson, L.A., Johnson, D.L. and Lampton, S.S. (Eds) (1983) Fisheries Techniques. American Fisheries Society, Bethseda.

Palmer, M.A., Bell, S.L. and Butterfield, I. (1992) A Botanical Classification of Standing Waters in Britain: Applications for Conservation and Monitoring, in Aquatic Conservation: Marine and Freshwater Ecosystems 2, 125-143, John Wiley & Sons Ltd, Chichester.

Parsons, M.S. (1993) A Review of the Scarce and Threatened Pyralid Moths of Great Britain. UK Conservation, No.11, Joint Nature Conservation Committee, Peterborough.

Perring, F.H. and Farrell, L. (1983) British Red Data Books: 1. Vascular Plants (second edition). The Royal Society for Nature Conservation, Lincoln.

Pond Action (1993) The Oxfordshire Pond Survey, Vol. 2, 196-199. Oxford Brookes University, Oxford.

Putman, R.J.(1984) Facts from Faeces. *Mammal Review*, 14, 79-97.

Ratcliffe, D.A. (1977) A Nature Conservation Review: Volume 1. Cambridge University Press, Cambridge.

Richardson, D.H.S. (1992) Pollution Monitoring with Lichens. Richmond Publishing, Slough.

Ricker, W.E. (Ed) (1968) Methods for Assessment for Fish Produce in Fresh Water. Blackwell Scientific Publications, Oxford.

Rodwell, J. (1991) British Plant Communities: Volume 1, Woodlands and Scrub. Cambridge University Press, Cambridge.

Rodwell, J. (1992a) British Plant Communities: Volume 2, Mires and Heaths. Cambridge University Press, Cambridge.

Rodwell, J. (1992b) British Plant Communities: Volume 3, Grassland and Montane Vegetation. Cambridge University Press, Cambridge.

Rodwell, J. (in press (a)) Aquatic Communities, Swamps, Tall-Herb Fens.

Rodwell, J. (in press (b)) Maritime and Weed Communities.

Round, F.E. (1993) A Review and Methods for the Use of Epilthic Diatoms for Detecting and Monitoring Changes in River Water Quality. HMSO, London.

Shirt, D.B. (1987) British Red Data Books: 2. Insects. Nature Conservancy Council, Peterborough.

Sneddon, P. and Randall, R.E. (1993) The Coastal Vegetated Shingle Structures of Great Britain. Main report. Joint Nature Conservation Committee, Peterborough.

Southern, H.N. (1973) A Yardstick for Measuring Populations of Small Rodents. *Mammal Review*, 3, 1-10.

Southwood, T.R.E. (1977) Ecological Methods with Particular Reference to Insect Populations. Chapman and Hall, London.

Spencer, J. and Kirby, K.J. (1992) An Inventory of Ancient Woodland for England and Wales. *Biological Conservation*, 62, 77-93.

Staines, B.W. and Ratcliffe, P.R. (1987) Estimating the Abundance of Red Deer (*Cervus elaphus* L.) and Roe Deer (*Capreolus capreolus* L.) and their Current Status in Britain. Symposium of the Zoological Society, London, 58,131-152.

Stewart, N.F. and Church, J.M. (1992) Red Data Book of Britain and Ireland: Stoneworts. Joint Nature Conservation Committee, Peterborough.

Stewart, N.F. and Church, J. (in prep.) Red Data Book: Lichens. Joint Nature Conservation Committee, Peterborough.

Stewart, N.F. and Church, J. (in press) Red Data Book: Bryophytes. Joint Nature Conservation Committee, Peterborough.

Stewart, A., Pearman, D.A. and Preston, C.D. (Eds) (in press). Scarce Plants in Britain. Joint Nature Conservation Committee, Peterborough.

Stoddart, D.M. (1970) Individual Range, Dispersion and Dispersal in a Population of Water Voles (*Arvicola terrestris* (L.)). *Journal of Animal Ecology*, 39,403-425.

Stone, C.J., Webb, A., Barton, C.,Ratcliffe, N.,Reed, T.C., Tasker, M.L. and Pienkowski, M.W. (in press) An Atlas of Seabird Distribution in North West European Waters. Joint Nature Conservation Committee, Peterborough.

Strachan, R., Birks, J.D.S., Chanin, P.R.S. and Jefferies, D.J. (1990) Otter Survey of England 1984-1986. Nature Conservancy Council, Peterborough.

Strachan, R. and Jefferies, D.J. (1993) The Water Vole *Arvicola terrestris* in Britain 1989-1990: Its Distribution and Changing Status. The Vincent Wildlife Trust, London.

Strachan, R., Jefferies, D.J. and Chanin, P.F.R. (in prep.) Pine Marten Survey of England and Wales 1987-1988. Joint Nature Conservation Committee, Peterborough.

Stroud, D.A., Mudge, G.P. and Pienkowski, M.W. (1990) Protecting Internationally Important Bird Sites. Nature Conservancy Council Peterborough.

Tapper, S. (1992) Game Heritage. An Ecological Analysis of Game Keeping Records. Game Conservancy.

Tasker, M.L., Jones, P.H., Dixon, T.J. and Blake, B.F. (1984) Counting Seabirds at Sea from Ships: a review of methods employed and a suggestion for a standardized approach. AUK, 101, 567-577.

Tasker, M.L., Webb, A., Hall, A.J., Pienkowski, M.W. and Langslow, D.R. (1987) Seabirds in the North Sea. Nature Conservancy Council, Peterborough.

Taylor, R.H. (1956) The use of pellet counts for estimating the density of populations of the wild rabbit, *Oryctolagus cuniculus* (L.). *New Zealand Journal of Science and Technology*, 38, 236-256.

Team, P.A. (in press) Freshwater Ecosystems. In P. Morris and R. Therivel. Methods of Environmental Assessment. UCL Press, London.

Thompson, P.M. and Harwood, J.(1990) Methods for Estimating the Population Size of Common Seals, *Phoca vitulina*. *Journal of Applied Ecology*, 27,924-938.

Trout, R.C., Tapper, S.C. and Herradine, J. (1986) Recent Trends in the Rabbit Population in Britain. *Mammal Review*, 16,117-123.

Twigg, G.I. (1975a) Catching Mammals. *Mammal Review*, 5, 83-100.

Twigg, G.I. (1975b) Marking mammals. *Mammal Review*, 5, 101-116.

Usher, M.B. (Ed) (1986) Wildlife Conservation Evaluation. Chapman and Hall, London.

Velander, K.A.(1983) Pine Marten Survey of Scotland, England and Wales 1980-1982. The Vincent Wildlife Trust, London.

Wallace, I.D. (1991) A Review of the Trichoptera of Great Britain. Research and Survey in Nature Conservation, No.32, Nature Conservancy Council, Peterborough.

Wallace, I.D., Wallace, B. and Philipson, G.N. (1990) A Key to the Case-bearing Caddis Larvae of Britain and Ireland. Freshwater Biological Association, Ambleside.

Walsh, F., Lee, N. and Wood, C.M. (1991) The Environmental Assessment of Opencast Coal Mines. Occasional Paper 28, Department of Planning and Landscape, University of Manchester, Manchester.

Ward, A.J., Thompson, D. and Hiby, A.R. (1987) Census techniques for grey seal populations. Symposium of the Zoological Society, London, 58,181-191.

84   Watson, A. and Hewson, R. (1973) Population Densities of Mountain Hares *Lepus timidus* in Western Scottish and Irish Moors and on Scottish hills. *Journal of Zoology*, 170,151-159.

Webb, A., Harrison, N.M., Leaper, G.M., Steele, R.D., Tasker, M.L. and Pienkowski, M.W. (1990) Seabird Distribution West of Britain. Nature Conservancy Council, Peterborough.

Whilde, A. (1993) Threatened Mammals, Birds, Amphibians and Fish in Ireland. Irish Red Data Book 2: Vertebrates. HMSO, Belfast.

Woodroffe, G., Lawton, J.H. and Davidson, W.L. (1990a) Patterns in the Production of Latrines by Water Voles and their Use as Indices of Abundance in Population Surveys. *Journal of Zoology*, 220,439-445.

Woodroffe G., Lawton J.H. and Davidson W.L. (1990b) The Impact of Feral Mink (*Mustela vison*) on Water Voles (*Arvicola terrestris*) in the North Yorkshire Moors National Park. *Biological Conservation*, 51,49-62.

Wright, J.F., Armitage, P.D. and Furse, M.T. (1989) Prediction of Invertebrate Communities using Stream Measurements. *Regulated Rivers, Research and Management*, 4, 147-155.

Wyatt, G. (1991a) A Review of Phase 1 Habitat Surveys in England. Nature Conservancy Council, Peterborough.

Wyatt, G. (1991b) A Review of Phase 1 Habitat Surveys in England:
      Volume 2 - North West Region
      Volume 3 - North East Region
      Volume 4 - West Midlands Region
      Volume 5 - East Midlands Region
      Volume 6 - East Anglia Region
      Volume 7 - South West Region
      Volume 8 - South Region
      Volume 9 - South East Region
Great Britain Nature Conservation Resource Survey Project No.4, Nature Conservancy Council, Peterborough.

Appendix One

The EA Process

The EA process

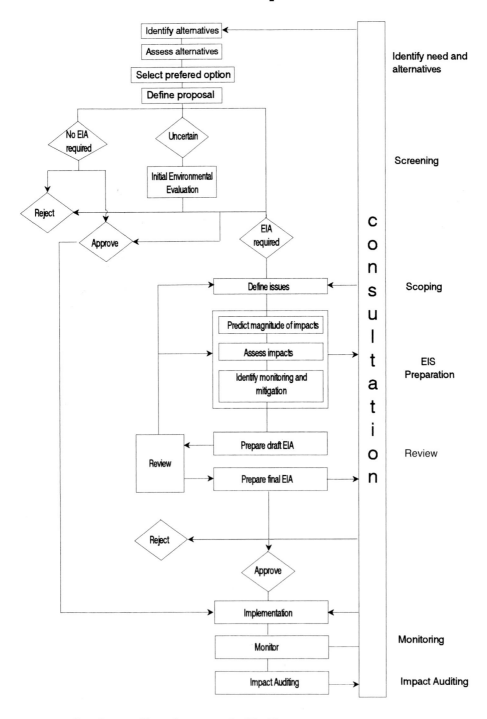

Flow diagram of the main components of the EA process
(Based on Wathern 1988)

Appendix Two

Consultees and Information Sources

Statutory Consultees

Local Authorities
English Nature/Scottish Natural Heritage/Countryside Council for
Wales/Environment Service - Countryside and Wildlife Branch
(Northern Ireland)
National Rivers Authority/Environmental Service - Environmental
Protection (Northern Ireland)/River Purification Boards (Scotland)
Countryside Commission
Ministry of Agriculture Fisheries and Food - Marine Environment
Protection Division
Scottish Agriculture and Fisheries Department
Welsh Office Agriculture Department

Other Possible Information Sources and Consultees

(For any EA only a small number of these organisations will need to
be contacted. In addition the list is not comprehensive and provides
only an indication of the range of possible information sources available.)

Assocation for Protection of Rural Scotland
Association of Drainage Authorities
Botanical Society of the British Isles (BSBI)
British Arachnological Society
British Association for Shooting and Conservation (BASC)
British Bryological Society
British Deer Society
British Dragonfly Society
Herpetological Conservation Trust
British Herpetological Society
British Lichen Society
British Mycological Society
British Phycological Society
British Recording in Scotland Committee (BRISC)
British Trust for Ornithology (BTO)
Butterfly Conservation
Conchological Society of Great Britain and Ireland
Council for Protection of Rural England (CPRE)
Fauna and Flora Preservation Society
Forestry Authority
Forestry Enterprise

Institute of Terrestrial Ecology
Joint Nature Conservation Committee (JNCC)
Ministry of Agriculture Fisheries and Food (MAFF)
MAFF Directorate of Fisheries Research
Marine Conservation Society
National Federation of Anglers
National Federation for Badger Groups
National Federation for Biological Recording
National Trust
Natural Environment Research Council (NERC)
Otter Project Wales
Plantlife
Port and Harbour Authorities
Royal Entomological Society
Royal Society for the Protection of Birds (RSPB)
Scottish River Purification Boards Association
Scottish Wildlife Trust
Sea Fisheries Committee (Association of England and Wales)
Sea Mammal Research Unit
Sea Watch Foundation
The Mammal Society
The Natural History Museum
The Otter Trust
The Wildlife Trusts
The World Conservation Union (IUCN)
Ulster Wildlife Trust
Vincent Wildlife Trust
Water Services (HQ) (Northern Ireland)
Whale and Dolphin Society
Wildfowl and Wetlands Trust
World Conservation Monitoring Centre (WCMC)
World Wide Fund for Nature (WWF)

Appendix Three

Protected Species in the UK

This appendix details species whose natural range includes the UK and which are protected under the:

- EC Directive on the Conservation of Natural Habitats and Wild Fauna and Flora (92/43/EEC) (EC Habitats Directive). The Habitats Directive aims to contribute to the maintenance of biodiversity by requiring Member States to take measures designed to maintain or restore certain natural habitats and wild species at a favourable conservation status within the Community. A series of sites, SACs, are to be designated for the purpose and strict protection of certain species is required. The Directive transposes the Bern Convention into community law.

- Convention on the Conservation of European Wildlife and Natural Heritage 1979 (Bern Convention). The 'Bern' Convention on the Conservation of European Wildlife and Natural Habitats encourages in particular the promotion of co-operation between countries in their conservation efforts, especially with regard to migratory species. Article 4(3) of the Convention states that parties should:

  "undertake to give special attention to the protection of areas that are of importance for the migratory species in Appendices II and III (including most birds) and which are appropriately situated in relation to migration routes as wintering, staging, feeding, breeding or moulting areas."

  Britain is a party to this convention, having ratified its provisions on 28 May 1982 (Stroud *et al* (1990))

- Convention on International Trade in Endangered Species (CITES). Species listed in Appendices I - III of CITES require an import/export licence to be moved into or out of the UK.

- Directive on the Conservation of Wild Birds (79/409/EEC) (EC Birds Directive). The Birds Directive provides for the protection, management and control of all species of naturally occurring wild birds. It requires Member States to take measures to maintain their populations of bird species at an ecologically and scientifically sound level and, for certain of those species, to classify the most suitable areas of habitat for them as SPAs.

• Wildlife and Countryside Act 1981 (Wildlife Order (Northern Ireland) 1985) and quinquennial review. According to Section 9 of the Wildlife and Countryside Act 1981, wild animals listed under Schedule 5 are protected from the following activities:

Part	1	Intentional killing, injuring and taking.
	2	Possession or control of any live or dead animal or any part of anything derived from such an animal.
	4a	Intentional damage to, destruction of or obstruction of access to any structure or place used for shelter or protection.
	4b	Disturbance of an animal occupying a structure or place used for shelter or protection.
	5a	Selling, offering for sale or in the possession or transport for the purpose of sale (this including any live or dead animal or any part or anything derived from such an animal).

According to Section 13 of the Wildlife and Countryside Act 1981, wild plants are protected from the following actions:

Part	1a	The intentional picking, uprooting or destruction of plants included in Schedule 8.
	1b	The unauthorised, intentional uprooting of any wild plant not included in Schedule 8.
	2a	The selling, offering for sale, possession or transport for the purpose of sale of any plant included in Schedule 8 (this including any live or dead wild plant, or any part of, or anything derived from such a plant).
	2b	The publishing of an advertisement for buying or selling any wild plant included in Schedule 8.

• Bonn Convention. The Convention on the Conservation of Migratory Species of Wild Animals (the Bonn Convention) is specifically concerned with migratory species. It provides for their conservation by giving strict protection to a number of endangered animals listed in its Appendix 1, whilst also providing the framework for a series of AGREEMENTS between Range States for the conservation and management of Appendix II species. Currently, an AGREEMENT on the conservation of migratory populations of Anatidae is being drawn up under the terms of the Convention. Britain is a party to this Convention, having signed it on 23 June 1979 and ratified its provisions on 23 July 1985 (Stroud et al (1990)).

- Protection of Badgers Act 1992. It is an offence to willfully take, kill, injure or illtreat a badger. Their setts are also protected against obstruction, destruction, or damage in any part, and the animals within a sett cannot be disturbed.

Tables summarising the species (other than birds) protected under the EC Habitats Directive, the Bern and Bonn Conventions, CITES, and the Wildlife and Countryside Act (Wildlife Order Northern Ireland) are presented as follows:

TABLE A   **Mammal species**

TABLE B   **Reptile species**

TABLE C   **Amphibian species**

TABLE D   **Fish species**

TABLE E   **Invertebrate species**

TABLE F   **Vascular plant species**

TABLE G   **Non-vascular plant species**

Bird species protected under Annex I of the EC Birds Directive, CITES and the Wildlife and Countryside Act (Wildlife Order Northern Ireland) are listed in Table H. This table only includes those species in categories A and C of the British list. Species covered in the Bern Convention have been incorporated into the above directives, conventions and legislation.

Species	EC Directive	Bern Convention	CITES	Wildlife and Countryside Act
	Annex(es)	Appendix	Appendix	Schedule
Natural Range UK				
Erinaceus europaeus (hedgehog)	-	III	-	6
Sorex araneus (common shrew)	-	III	-	6
S.minutus (pygmy shrew)	-	III	-	6
Neomys fodiens (water shrew)	-	III	-	6
Crocidura suaveolens (lesser white-toothed shrew) (Isles of Scilly only)	-	III	-	6
Rhinolophus ferrumequinum (greater horseshoe bat)	IIa,IVa	II	-	5,6
R.hipposideros (lesser horseshoe bat)	IIa,IVa	II	-	5,6
Myotis mystacinus (whiskered bat)	IVa	II	-	5,6
M.brandtii (Brandt's bat)	IVa	II	-	5,6
M.nattereri (Natterer's bat)	IVa	II	-	5,6
M.bechsteini (Bechstein's bat)	IIa,IVa	II	-	5,6

M.myotis (mouse-eared bat) (probably extinct)	IIa,IVa	II	-	5,6
M.daubentonii (Daubenton's bat)	IVa	II	-	5,6
Eptesicus serotinus (serotine)	IVa	II	-	5,6
Nyctalus noctula (noctule)	IVa	II	-	5,6
N.leisleri (Leisler's bat)	IVa	II	-	5,6
Pipistrellus pipistrellus (pipistrelle)	IVa	III	-	5,6
Barbastella barbastellus (barbastelle)	IIa,IVa	II	-	5,6
Plecotus auritus (brown long-eared bat)	IVa	II	-	5,6
P.austriacus (grey long-eared bat)	IVa	II	-	5,6
Muscardinus avellanarius (dormouse)	IVa	III	-	5,6
Sciurus vulgaris (red squirrel)	-	III	-	5,6
Lepus timidus (mountain hare)	Va	III	-	-
Balaenoptera physalus (fin whale)	IVa	III	I	5
B.acutorostrata (Minke whale)	IVa	III	I	5

Orcinus orca (killer whale)	IVa	II	II	5
Grampus griseus (Risso's dolphin)	IVa	II	II	5
Globicephala melaena (long-finned pilot whale)	IVa	II	II	5
Delphinus delphis (common dolphin)	IVa	II	II	5,6
Tursiops truncatus (tursio) (bottle-nosed dolphin)	IIa,IVa	II	II	5,6
Lagenorhynchus acutus (Atlantic white-sided dolphin)	IVa	II	II	5
L.albirostris (white-beaked dolphin)	IVa	II	II	5
Stenella coerulealba (striped dolphin)	IVa	II	II	5
Phocena phocena (harbour porpoise)	IIa,IVa	II	II	5,6
Mesoplodon bidens (Sowerby's beaked whale)	IVa	II	II	5
Lutra lutra (otter)	IIa,IVa	II	I	5,6
Meles meles (badger)	-	III	-	6 (Badgers Act)
Mustela erminea (stoat)	-	III	III	- (UK reservation)
M.nivalis (weasel)	-	III	-	-
M.(Putorius) putorius (polecat)	Va	III	-	6

Martes martes (pine marten)	Va	III	-	5,6
Felis silvestris (catus) (wildcat)	IVa	II	II	5,6
Phoca vitulina (common seal)	IIa	III	-	(Seals Act)
Halichoerus grypus (grey seal)	IIa	III	-	(Seals Act)
Cervus elaphus (red deer)	-	III	-	(Deer Act)
Capreolus capreolus (roe deer)	-	III	_	(Deer Act)

Probably Vagrant

Vespertillio murinus (parti-coloured bat)	IVa	II	-	5,6
Eptesicus nilssonii (northern bat)	IVa	II	-	5,6
Pipistrellus nathusii (Nathusius pipistrelle)	IVa	II	-	5,6
Pseudorca crassidens (false killer whale)	IVa	II	II	5
Hyperoodon ampullatus (northern bottlenose whale)	IVa	III	I	5
Mesoplodon mirus (True's beaked whale)	IVa	II	II	5
M.europaeus (Gearvais' beaked whale)	IVa	III	II	5
Ziphius cavirostris (Cuvier's beaked whale)	IVa	II	II	5

Kogia (Physeter) breviceps (pygmy sperm whale)	IVa	III	II	5
Physeter macrocephalus (sperm whale)	IVa	III	I	5
Balaenoptera (Sibbaldus) musculus (blue whale)	IVa	II	I	5
B.borealis (sei whale)	IVa	III	I	5
Megaptera novaeangliae (humpback whale)	IVa	II	I	5
Delphinapterus leucas (white whale)	IVa	III	II	5
Monodon nonoceros (narwhal)	IVa	III	II	5
Eubalaena glacialis (northern right whale)	IVa	II	I	5
Phoca (Pusa) hispida (ringed seal)	Va	III	-	-
Phoca groenlandica (Pagophilus groenlanicus) (harp seal)	Va	III	-	-
Erignathus barbatus (bearded seal)	Va	III	-	-
Cystophora cristata (hooded seal)	Va	III	-	-
Odobenus rosmarus (walrus)	-	II	III	5

Established Aliens

Glis glis (fat dormouse)	-	III	-	6,9

Hystrix cristata (European porcupine)	IVa	II	-	9
Dama dama (fallow dear) (ancient introduction)	-	III	-	(Deer Act)
Rangifer tarandus (reindeer)	-	III	-	-
Other Cervidae (other deer)	-	III	-	(Sika included in Deer Act)
Capra aegagrus (bezoar goat)	IIa (natural populations only)	II	-	-

Table B - reptile species occurring in the UK and listed in the EC Habitats Directive, the Bern Convention, CITES and the Wildlife and Countryside Act.

Species	EC Directive	Bern Convention	CITES	Wildlife and Countryside Act
	Annex(es)	Appendix	Appendix	Schedule
Natural Range UK				
Lacerta agilis (sand lizard)	IVa	II	-	5
L.vivipara (viviparous lizard)	-	III	-	5
Anguis fragilis (slow worm)	-	III	-	5
Natrix natrix (grass snake)	-	III	-	5
Coronella austriaca (smooth snake)	IVa	II	-	5
Vipera berus (adder)	-	III	-	5
Vagrants				
Dermochelys coriacea (leatherback turtle)	IVa	II	I	5
Caretta caretta (loggerhead turtle)	*IIa, IVa	III	I	5
Lepidochelys kempii (Kemp's Ridley turtle)	IVa	II	I	
Chelonia mydas	IVa	II	I	5

(green turtle)

Eretmochelys imbricata (hawksbill turtle)	IVa	II	I	5

Recently Established
Aliens

Emys orbicularis (European pond terrapin)	IIa, IVa	II	-	9
Podarcis muralis (common wall lizard)	IVa	II	-	9
Elaphe longissima (Aesculapean snake)	IVa	II	-	9

Note

* Priority species

Table C - amphibian species occurring in the UK and listed in the EC Habitats Directive, the Bern Convention, CITES and the Wildlife and Countryside Act.

Species	EC Directive	Bern	CITES	Wildlife and Countryside Act
	Annex(es)	Appendix	Appendix	Schedule
Natural Range UK				
Triturus cristatus (great crested newt)	IIa, IVa	II	-	5
T. vulgaris (smooth newt)	-	III	-	5
T. helveticus (palmate newt)	_	III	-	5
Rana temporaria (common frog)	Va	III	-	5
Bufo bufo (common toad)	-	III	-	5
B. calamita (natterjack toad)	IVa	II	-	5
Recently Established Aliens				
Triturus alpestris (alpine newt)	-	III	-	9
T. carnifex (Italian crested newt)	IVa	II	-	9
Bombina variegata (yellow-bellied toad)	IVa	II	-	9
Alytes obstetricans (midwife toad)	IVa	II	-	9

Rana esculenta (edible frog)	Va	III	-	9
R.ridibunda (marsh frog)	Va	III	-	9
R.lessonae (pool frog) (some populations possibly native)	IVa	III	-	-
Hyla arborea (European tree frog) (some populations possibly native)	IVa	II	-	9

Table D - fish species occurring in the UK and listed in the EC Habitats Directive, the Bern Convention, CITES and the Wildlife and Countryside Act.

Species	EC Directive	Bern Convention	CITES	Wildlife and Countryside Act
	Annex(es)	Appendix	Appendix	Schedule
Natural Range UK				
Lampetra fluviatilis (river lamprey)	IIa, Va	III	-	-
L.planeri (brook lamprey)	IIa	III	-	-
Petromyzon marinus (sea lamprey)	IIa	III	-	-
Alosa alosa (allis shad)	IIa, Va	III	-	5
A.fallax (twaite shad)	IIa, Va	III	-	-
Coregonus albula (vendace)	Va	III	-	5
C.lavaretus (whitefish)	Va	III	-	5
Thymallus thymallus (grayling)	Va	III	-	-
Salmo salar (Atlantic salmon)	IIa, Va (in freshwater only)	III	-	-
Barbus barbus (barbel)	Va	-	-	-
Cobitis taenia (spined loach)	IIa	III	-	-

Potamoschistus microps (common goby)	-	III	-	-
P.minutus (sand goby)	-	III	-	-
Cottus gobio (bullhead)	IIa	-	-	-

Vagrant

Acipenser sturio (sturgeon)	*IIa, IVa	III	I	5

Believed Extinct

Coregonus oxyrinchus (houting)(anadromous populations only)	*IIa, IVa	III	-	—

Introduced Aliens

Rhodeus sericeus (bitterling)	IIa	III	-	-
Siluris glanis (wels)	-	III	-	9

Note

* Priority species

Table E - invertebrate species occurring in the UK and listed in the EC Habitats Directive, the Bern Convention, CITES and the Wildlife and Countryside Act.

Species	EC Directive	Bern Convention	CITES	Wildlife and Countryside Act
	Annex(es)	Appendix	Appendix	Schedule
Natural Range UK				
Austropotamobius pallipes (Atlantic stream crayfish)	IIa, Va	III	-	5
Lucanus cervus (stag beetle)	IIa	III	-	-
Limoniscus violaceus (violet click beetle)	IIa	-	-	5
Eurodryas(Euphydryas) aurinia (marsh fritillary butterfly)	IIa	II	-	5
Callimorpha quadripunctata (Euplagia quadripunctaria) (Jersey tiger moth)	*IIa	-	-	-
Coenagrion mercuriale (southern damselfly)	IIa	II	-	-
Margaritifera margaritifera (pearl mussel)	IIa,V	III	-	5
Vertigo angustior (snail)	IIa	-	-	-
V.genesii (snail)	IIa	-	-	-
V.geyeri (snail)	IIa	-	-	-
V.moulinsiana (snail)	IIa	-	-	-

Hirudo medicinalis (medicinal leech)	Va	III	II	5

Extinct

Oxygastra curtisii (orange-spotted emerald dragonfly)	IIa,IVa	II	-	-
Graphoderus bilineatus (water beetle)	IIa,IVa	II	-	-
Cerambyx cerdo (beetle)	IIa,IVa	II	-	-

Extinct but
Re-established

Lycaena dispar (large copper butterfly)	IIa,IVa	II	-	5
Maculinea arion (large blue butterfly)	IVa	II	-	5

Vagrants

Parnassius apollo (Apollo butterfly)	IVa	-	II	-
Proserpinus proserpina (Curzon's sphinx moth)	IVa	II	-	-

Established Aliens

Helix pomatia (Roman snail) (probably ancient introduction)	Va	III	-	-
Astacus astacus (noble crayfish)	Va	III	-	9

Table F - vascular plant species occurring in the UK and listed in the EC Habitats Directive, the Bern Convention, CITES and the Wildlife and Countryside Act.

Species	EC Directive	Bern Convention	CITES	Wildlife and Countryside Act
	Annex(es)	Appendix	Appendix	Schedule
Natural Range UK				
Lycopodium - all species (clubmosses)	Vb	-	-	-
Trichomanes speciosum (Killarney fern)	IIb,IVb	I	-	8
Apium repens (creeping marshwort)	IIb,IVb	I	-	8
Cypripedium calceolus (lady's slipper)	IIb,IVb	I	II	8
Gentianella anglica (early gentian)	IIb,IVb	I	-	8
Liparis loeselii (fen orchid)	IIb,IVb	I	II	8
Luronium natans (floating-leaved water plantain)	IIb,IVb	I	-	8
Najas flexilis (slender naiad)	IIb,IVb	I	-	8
Rumex rupestris (shore dock)	IIb,IVb	I	-	8
Ruscus aculeatus (butcher's broom)	Vb	-	-	-
Saxifraga hirculus (yellow marsh saxifrage)	IIb,IVb	I	-	8

All Orchidaceae	**	**	II	11 species on 8

Believed Extinct

Bromus interruptus (interrupted brome grass)	-	I	-	-
Spiranthes aestivalis (summer lady's tresses)	IIb,IVb	I	II	-

**Probably Established
Aliens**

Galanthus nivalis (snowdrop)(may be native in a few places in Wales and W England)	Vb (if native)	-	II	-

Note

** See *Cypripedium calceolus, Liparis loeselii, Spiranthes aestivalis.* Orchid-rich dry calcareous grasslands are listed on the EC Habitats Directive Annex I as a priority habitat requiring the designation of protected areas.

Table G - non-vascular plant species occurring in the UK and listed in the EC Habitats Directive, The Bern Convention, CITES and the Wildlife and Countryside Act.

Species	EC Directive	Bern Convention	CITES	Wildlife and Countryside Act
Natural Range UK	Annex(es)	Appendix	Appendix	Schedule
Lithothamnium corallioides (maerl)	Vb	-	-	-
Phymatolithon calcareum (maerl)	Vb	-	-	-
Marsupella profunda (liverwort)	*IIb	I	-	8
Petalophyllum ralfsii (liverwort)	IIb	I	-	8
Buxbaumia viridis (moss)	IIb	I	-	8
Drepanocladus vernicosus (moss)	IIb	I	-	8
Leucobryum glaucum (moss)	Vb	-	-	-
Sphagnum - all species (bog mosses)	Vb	-	-	8 (*Sphagnum balticum* only)
Cladonia sub-genus *Cladina* - all species	Vb	-	-	-

Note

*   Priority species

Table H - bird species appearing in the UK and listed in the EC Birds Directive, CITES and the Wildlife and Countryside Act

Species	EC Birds Directive	CITES Appendix	Wildlife and Countryside Act Schedule
Gavia stellata (red-throated diver)	1		1.1,4
Gavia arctica (black-throated diver)	1		1.1,4
Gavia immer (great northern diver)	1		1.1,4
Gavia adamsii (white-billed diver)			1.1,4
Podiceps auritus (Slavonian grebe)	1		1.1,4
Podiceps nigricollis (black-necked grebe)			1.1,4
Calonectris diomedea (Cory's shearwater)	1		
Puffinus assimilis (little shearwater)	1		
Hydrobates pelagicus (storm petrel)	1		
Oceanodroma leucorhoa (Leach's petrel)	1		1.1,4
Phalacrocorax carbo (cormorant)	1 *(sinensis)*		
Botaurus stellaris (bittern)	1		1.1,4
Ixobrychus minutus (little bittern)	1		1.1,4

Nycticorax nycticorax (night heron)	1		
Ardeola ralloides (squacco heron)	1		
Bubulcus ibis (cattle egret)		III	
Egretta garzetta (little egret)	1	III	
Egretta alba (great white egret)	1	III	
Ardea purpurea (purple heron)	1		1.1
Ciconia nigra (black stork)	1		
Ciconia ciconia (white stork)	1		
Plegadis falcinellus (glossy ibis)	1		
Platalea leucorodia (spoonbill)	1	II	1.1,4
Cygnus columbianus bewickii (Bewick's swan)	1		1.1
Cygnus cygnus (whooper swan)	1		1.1
Anser brachyrhynchus (pink-footed goose)			2.1
Anser albifrons flavirostris (white-fronted goose)	1		2.1
Anser erythropus (lesser white-fronted goose)	1		

Anser anser (greylag goose)		1.II, 2.1 (restricted areas)
Branta canadensis (Canada goose)		2.1
Branta leucopsis (barnacle goose)	1	
Branta ruficollis (red-breasted goose)	1	
Alopochen aegyptiacus (Egyptian goose)	III	
Anas penelope (wigeon)	III	2.1
Anas strepera (gadwall)		2.1
Anas formosa (Baikal teal)	II	
Anas crecca (teal)	III	2.1
Anas platyrhynchos (mallard)		2.1
Anas acuta (pintail)	III	1.II, 2.1
Anas querquedula (garganey)	III	1.II
Anas clypeata (shoveler)	III	2.1
Aythya ferina (pochard)		2.1
Aythya nyroca (ferruginous duck)	III	

Aythya fuligula (tufted duck)			2.1
Aythya marila (scaup)			1.1
Clangula hyemalis (long-tailed duck)			1.1,4
Melanitta nigra (common scoter)			1.1,4
Melanitta fusca (velvet scoter)			1.1,4
Bucephala clangula (goldeneye)	1		1.II
Pernis apivorus (honey buzzard)	1		1.1,4
Milvus migrans (black kite)	1		4
Milvus milvus (red kite)	1		1.1,4
Haliaeetus albicilla (white-tailed eagle)	1	I	1.1,4
Circus aeruginosus (marsh harrier)	1		1.1,4
Circus cyaneus (hen harrier)	1		1.1,4
Circus pygargus (Montagu's harrier)	1		1.1,4
Accipiter gentilis (goshawk)			1.1,4
Accipiter nisus (sparrowhawk)			4

Buteo buteo (buzzard)		4
Buteo lagopus (rough-legged buzzard)		4
Aquila chrysaetos (golden eagle)	1	1.1,4
Pandion haliaetus (osprey)	1	1.1,4
Falco naumanni (lesser kestrel)		4
Falco tinnunculus (kestrel)		4
Falco sparverius (American kestrel)		4
Falco vespertinus (red-footed falcon)		4
Falco columbarius (merlin)	1	1.1,4
Falco subbuteo (hobby)		1.1,4
Falco eleonorae (Eleonora's falcon)		4
Falco rusticolus (gyrfalcon)	1	1.1,4
Falco peregrinus (peregrine)	1	1.1,4
Tetrao urogallus (capercaillie)	1	1
Coturnix coturnix (quail)		1.1,4

Porzana porzana (spotted crake)	1		1.1,4
Porzana parva (little crake)	1		
Porzana pusilla (Baillon's crake)	1		
Crex crex (corncrake)	1		1.1,4
Gallinula chloropus (moorhen)			2.1
Fulica atra (coot)			2.1
Grus grus (crane)	1	II	
Grus canadensis (sandhill crane)		II	
Tetrax tetrax (little bustard)	1		
Chlamydotis undulata (houbara)	1		
Otis tarda (great bustard)	1	II	
Himantopus himantopus (black-winged stilt)	1		1.1,4
Recurvirostra avosetta (avocet)	1		1.1,4
Burhinus oedicnemus (stone-curlew)	1		1.1,4
Cursorius cursor (cream-coloured courser)	1		

Glareola pratincola (collared pratincole)	1	
Charadrius dubius (little ringed plover)		1.1,4
Charadrius alexandrinus (Kentish plover)		1.1,4
Charadrius morinellus (dotterel)	1	1.1,4
Pluvialis apricaria (golden plover)	1	2.1
Calidris temminckii (Temminck's stint)		1.1,4
Calidris maritima (purple sandpiper)		1.1,4
Philomachus pugnax (ruff)	1	1.1,4
Gallinago gallinago (snipe)		2.1
Gallinago media (great snipe)	1	
Scolopax rusticola (woodcock)		2.1
Limosa limosa (black-tailed godwit)		1.1,4
Numenius phaeopus (whimbrel)		1.1,4
Tringa nebularia (greenshank)		1.1,4
Tringa ochropus (green sandpiper)		1.1,4

Tringa glareola (wood sandpiper)	1	1.1,4
Phalaropus lobatus (red-necked phalarope)	1	1.1,4
Larus melanocephalus (Mediterranean gull)	1	1.1
Larus minutus (little gull)		1.1
Larus genei (slender-billed gull)	1	
Gelochelidon nilotica (gull-billed tern)	1	
Sterna caspia (Caspian tern)	1	
Sterna sandvicensis (sandwich tern)	1	
Sterna dougallii (roseate tern)	1	1.1,4
Sterna hirundo (common tern)	1	
Sterna paradisaea (Arctic tern)	1	
Sterna albifrons (little tern)	1	1.1,4
Chlidonias hybridus (whiskered tern)	1	
Chlidonias niger (black tern)	1	1.1,4
Columba livia (rock dove)	III	

Streptopelia turtur (turtle dove)		III	
Psittacula krameri (ring-necked parakeet)		III	
Tyto alba (barn owl)			1.1,99
Nyctea scandiaca (snowy owl)	1		1.1
Asio flammeus (short-eared owl)	1		
Aegolius funereus (Tengmalm's owl)	1		
Caprimulgus europaeus (nightjar)	1		
Alcedo atthis (kingfisher)	1		1.1,4
Merops apiaster (bee-eater)			1.1,49
Coracias garrulus (roller)	1		
Upupa epops (hoopoe)			1.1,4
Jynx torquilla (wryneck)			1.1,4
Melanocorypha calandra (calandra lark)	1		
Calandrella brachydactyla (short-toed lark)	1		
Lullula arborea (wood lark)	1		1.1,4

Eremophila alpestris (shore lark)		1.1,4
Anthus campestris (tawny pipit)	1	
Luscinia svecica (bluethroat)	1	1.1,4
Phoenicurus ochruros (black redstart)		1.1,4
Turdus pilaris (fieldfare)		1.1,4
Turdus iliacus (redwing)		1.1,4
Cettia cetti (Cetti's warbler)		1.1,4
Locustella luscinioides (Savi's warbler)		1.1,4
Acrocephalus melanopogon (moustached warbler)	1	
Acrocephalus paludicola (aquatic warbler)	1	
Acrocephalus palustris (marsh warbler)		1.1,4
Sylvia sarda (Marmora's warbler)	1	
Sylvia undata (Dartford warbler)	1	1.1,4
Sylvia rueppelli (Ruppell's warbler)	1	
Sylvia nisoria (barred warbler)	1	

Regulus ignicapillus (firecrest)		1.1,4
Ficedula parva (red-breasted flycatcher)	1	
Ficedula albicollis (collared flycatcher)	1	
Panurus biarmicus (bearded tit)		1.1,4
Parus cristatus (crested tit)		1.1,4
Certhia brachydactyla (short-toed treecreeper)		1.1,4
Oriolus oriolus (golden oriole)		1.1,4
Lanius collurio (red-backed shrike)	1	1.1,4
Lanius minor (lesser grey shrike)	1	
Pyrrhocorax pyrrhocorax (chough)	1	1.1,4
Fringilla montifringilla (brambling)		1.1
Serinus serinus (serin)		1.1,4
Loxia leucoptera (two-barred crossbill)		1.1,4
Loxia curvirostra (crossbill)		1.1,4
Loxia scotica (Scottish crossbill)	1	1.1,4

Loxia pytopsittacus (parrot crossbill)		1.1,4
Carpodacus erythrinus (scarlet rosefinch)		1.1,4
Calcarius lapponicus (Lapland bunting)		1.1,4
Plectrophenax nivalis (snow bunting)		1.1,4
Emberiza cirlus (cirl bunting)		1.1,4
Emberiza hortulana (ortolan bunting)	1	
Emberiza caesia (Cretzschmar's bunting)	1	

Appendix Four

Licensing Procedures for Protected Species

All field surveyors must ensure that they follow the correct licensing procedures when conducting studies of protected species. Licences are required in the following situations:

Under the Wildlife and Countryside Act 1981

- To kill, to take or have in possession any wild animal included in Schedule 5 of the Act, or to have in possession any part or derivative of such an animal.

- To take and possess for the purposes of photography any wild animal included in Schedule 5.

- To kill or take any wild animal included in Schedule 6 of the Act by any method prohibited in Section 11 of the Act.

- To examine the nests of, or to disturb for the purposes of photography any wild bird included in Schedule 1 of the Act.

- To take wild birds for the purpose of ringing or marking or examining any ring or mark.

- To kill, take or have in possession wild birds, or to take or have in possession wild birds' eggs.

- To damage, destroy or obstruct access to any structure or place which any wild animal included in Schedule 5 uses for shelter or protection; or to disturb any such animal when it is occupying a structure or place which it uses for shelter or protection.

- To uproot any wild plant without the landowner's permission.

- To pick, uproot or destroy any wild plant listed in Schedule 8 of the Act, or any wild plant not included in that Schedule.

Under the Protection of Badgers Act 1992

- To kill, injure or take badgers, or to possess badgers or any part of a badger for scientific or educational purposes, for the purpose of any zoological collection, or for the purpose of marking.

- To interfere with, or destroy, a badger sett for the purpose of development of land, preserving or archaeological investigation of an ancient monument, for investigation of offenses at setts, or for the control of foxes for the protection of wildlife.

Further information and licensing forms can be obtained from:

Licensing Section
English Nature
Northminster House
PETERBOROUGH
PE1 1UA
Tel: 01733 340345

Licensing Section
Countryside Council for Wales
Plas Penrhos
Penrhos Road
BANGOR
Gwynedd
LL57 2LQ
Tel: 01248 370444

Licensing Section
Scottish Natural Heritage
2/5 Anderson Place
Bonnington Bond
EDINBURGH
EH6 5NP
Tel: 0131 554 9797

Further information on licensing requirements in Northern Ireland under the Wildlife Order 1985 can be obtained from:

Department of the Environment - Environment Service
Countryside and Wildlife Branch
Commonwealth House
Castle Street
Belfast BT1 1FY
Tel: 01232 311808

Appendix Five

Annex 1 Habitat Types found in the UK

(+ = Priority Habitat type)

Coastal and halophytic habitats

Open sea and tidal areas

11.25		Sandbanks which are slightly covered by seawater all the time
13.2		Estuaries
14		Mudflats and sandflats not covered by seawater at low tide
21	+	Lagoons
		Large shallow inlets and bays
		Reefs

Sea cliffs and shingle or stony beaches

17.2	Annual vegetation of drift lines
17.3	Perennial vegetation of stony banks
18.21	Vegetated sea cliffs of the Atlantic and Baltic coasts

Atlantic and continental salt marshes and salt meadows

15.11		*Salicornia* and other annuals colonising mud and sand
15.12		*Spartina* swards (*Spartinion*)
15.13		Atlantic salt meadows (*Glauco-Puccinellietalia*)
15.14	+	Continental salt meadows (*Puccinellietalia distantis*)

Mediterranean and thermo-Atlantic salt marshes and salt meadows

15.15	Mediterranean salt meadows (*Juncetalia maritimi*)
15.16	Mediterranean and thermo-Atlantic halophilous scrubs (*Arthrocnemetalia fructicosae*)

Coastal sand dunes and continental dunes

Sea dunes of the Atlantic, North Sea and Baltic coasts

16.211	Embryonic shifting dunes

16.212	Shifting dunes along the shoreline with *Ammophilia arenaria* (white dunes)
16.221-	
16.22+	Fixed dunes with herbaceous vegetation (grey dunes):
(16.221)	*Galio-koelerion albescentis*
(16.225)	*Mesobromion*
(16.226)	*Trifolio-Gerantietea sanguinei, Galio maritimi-Geranion sanguinei*
(16.227)	*Thero-Airion, Botrychio-Polygaletum, Tuberarion guttatae*
16.23 +	Decalcified fixed dunes with *Empetrum nigrum*
16.24 +	Eu-Atlantic decalcified fixed dunes (*Calluno-Ulicetea*)
16.25	Dunes with *Hippophae rhamnoides*
16.26	Dunes with *Salix arenaria*
16.27 +	Dune juniper thickets (*Juniperus* spp)
16.29	Wooded dunes of the Atlantic coast
16.31-16.35	Humid dune slacks
1.A	Machairs (& machairs in Ireland)

Continental dunes, old and decalcified

| 64.1 & 34.2 | Open grassland with *Corynephorus* and *Agrostis* of continental dunes |

Freshwater habitats

Standing water

22.11&22.31	Oligotrophic waters containing very few minerals of Atlantic sandy plains with amphibious vegetation: *Lobelia, Littorelia* and *Isoetes*.
22.12&(22.31 &22.32)	Oligotrophic waters in medio-European and perialpine area with amphibious vegetation: *Littorella* or *Isoetes* or annual vegetation on exposed banks (*Nanocyperetalia*)
21.12&22.44	Hard oligo-mesotrophic waters with benthic vegetation of *Chara* formations
22.13	Natural eutrophic lakes with *Magnopotamion* or *Hydrocharition*-type vegetation
22.14	Dystrophic lakes
22.34 +	Mediterranean temporary ponds

24.4   Floating vegetation of *Ranunculus* of plane and sub-mountainous rivers

Temperate heath and scrub

31.11   Northern Atlantic wet heaths with *Erica tetralix*
31.12+   Southern Atlantic wet heaths with *Erica Ciliaris* and *Erica tetralix*
31.2   Dry heaths (all sub-types)
31.234   +   Dry coastal heaths with *Erica vagans* and *Ulex maritimus*
31.4   Alpine and sub-alpine heaths
31.622   Sub-Arctic willow scrub

Sclerophylous scrub (Matorral)

Sub-Mediterranean and temperate

31.82   Stable *Buxus sempervirens* formations on calcareous rock slopes (*Berberidion* p.)
31.88   *Juniperus communis* formations on calcareous heaths or grasslands

Natural and semi-natural grassland formations

Natural grasslands

34.2   Calaminarian grasslands
36.32   Siliceous alpine and boreal grass
36.41 to
36.45   Alpine calcareous grasslands

Semi-natural dry grasslands and scrubland facies

34.31 to
34.34   On calcareous substrates (*Festuco-Brometalia*)
34.31 to
34.34   +   On calcareous substrates (*Festuco-Brometalia*) (+ important orchid sites)
35.1   +   Species-rich *Nardus* grassland, on siliceous substrates in mountain areas (and sub-mountain areas, in continental Europe)

| 37.31 | *Molinia* meadows on chalk and clay (eu-molinion) |
| 37.7 & 37.8 | Eutrophic tall herbs |

Mesophile grasslands

| 38.2 | Lowland hay meadows (*Alopecurus pratensis, Sanguisorba officinalis*) |
| 38.3 | Mountain hay meadows (British types with *Geranium sylvaticum*) |

Raised bogs, mires and fens

Sphagnum acid bogs

51.1 +	Active raised bogs
51.2	Degraded raised bogs (still capable of natural regeneration)
52.1 & 52.2 +	Blanket bog (+active only)
52.1 & 52.2	Blanket bog
54.5	Transition mires and quaking bogs
54.6	Depressions on peat substrates (*Rynchosporion*)

Calcareous fens

53.3+	Calcareous fens with *Cladium mariscus* and *Carex davailliana*
54.12 +	Petrifying springs with tufa formation (*Cratoneurion*)
54.2	Alkaline fens
54.3+	Alpine pioneer formations of *Caricion bicolaris-atrofuscae*

Rocky habitats and caves

Scree

| 6.1 | Siliceous |
| 61.2 | Eutric |

62.1 &
62.1A   Calcareous sub-types
62.2   Silicicolous sub-types
62.3   Pioneer vegetation of rock surfaces
62.4   +   Limestone pavements

Other rocky habitats

65   Caves not open to the public

Forests

Forests of temperate Europe

41.12   Beech forests with *Ilex* and *Taxus* rich in epiphytes
               (*Illici fagion*)
41.13   *Asperulo-Fagetum* beech forests
41.24   *Stellario-Carpinetum* oak-hornbeam forests
41.4   +   *Tilio-Acerion* ravine forests
41.51   Old acidophilous oak woods with *Quercus robur* on
               sandy plains
41.53   Old oak woods with *Ilex* and *Blechnum* in the British
               Isles
42.51   +   Caledonian forest
44.A1 -
44.A4   +   Bog woodland
44.3   +   Residual alluvial forests (*Alnion glutinoso-incanae*)

Mediterranean sclerophyllous forests

Mediterranean mountainous coniferous forests

42.A71 -
42.A73 +   *Taxus baccata* woods

Appendix Six

List of Professional Institutes

Institute of Biology
20 Queensbury Place
London
SW7 2DZ

Tel: 0171 581 8333
Fax: 0171 823 9409

Institute of Ecology and Environmental Management (IEEM)
36 Kingfisher Court
Hambridge Road
Newbury
RG14 5SJ

Tel: 01635 37715
Fax: 01635 550230

Institute of Environmental Sciences
14 Princes Gate
London
SW7 1PU
Tel: 01778 394846

Institute of Fisheries Management
Balmaha
Coldwells Road, Holmer
Hereford

Tel: 01432 276225

Institute of Water and Environmental Management
15 John Street
London
WC1N 2EB

Tel: 0171 831 3110
Fax: 0171 405 4967

Index